BIOCATALYSIS AND BIOCATALYTIC TECHNOLOGIES

Biocatalysis and Biocatalytic Technologies

G. E. Zaikov
Editor

Nova Science Publishers, Inc.
New York

For permission to use material from this book please contact us:
Telephone 631-231-7269; Fax 631-231-8175
Web Site: http://www.novapublishers.com

Library of Congress Cataloging-in-Publication Data
Available upon request.

ISBN 1-60021-041-4

Published by Nova Science Publishers, Inc. ✦ New York

Contents

Preface

This new book covers the application, both actual and potential, of biological catalysts (including whole cells or isolated components thereof, natural and modified enzymes and catalytic antibodies) for the synthesis, interconversion or degradation of chemical species.

Chapter one investigates nanoparticles, possessed by low Curie temperature, which can be used as self-controlled mediators in the treatment of tumours by a method of magnetic hyperthermia. Authors carried out the synthesis of nanoparticles, made the physical and chemical analysis, and were offered the device, allowing qualitatively and quantitatively to estimate efficiency of a method of magnetic hyperthermia.

In chapter two, the phenylalanine – ammonia-lyase (PAL) of *Rhodotorula aurantiaca* KM-1 was isolated and purified up to homogeneity. The molecular weight of this tetramer enzyme was 291 kDa, and it comprised homologue subunits with molecular weight of 80 kDa. The isoelectric point of the enzyme was reached at pH 5,3. Polyclonal rabbit antibodies to PAL were obtained. The enzyme expressed maximal activity in reaction to de-amination under $50^{\circ}C$ and pH 8,2. The enzyme deaminated L-phenylalanine and L-tyrosine. D-phenylalanine, L-tryptophan and L-dioxyphenylalanine were not substrates for this enzyme.

The construction of the immobilized biocatalytic system on the basis of artificial microbial consortium for parathion biodegradation was carried out in chapter three. Some conditions of coexistence of various cultures in a joint system were investigated. The positive influence of several microorganisms (*E. coli* DH5α, *Pseudomonas fluorescens* B-473, *Pseudomonas fluorescens* B-495) on the *p*-NP degradation by *Pseudomonas putida* cells were revealed. All these cultures were introduced into the joint microbial system and the optimized biomass content of the consortium was established. The entrapment of the developed complex biocatalytic system into the poly(vinyl alcohol) cryogel allowed for the creation of immobilized consortium. This immobilized biocatalyst was used for the biodegradation of PT and *p*-NP under the batch conditions and obvious advantages compared to the immobilized single culture *Ps. putida*.

To determine the total concentration of cells composing the corrosive biofilms, a new approach was developed using bioluminescent method of ATP detection. The express screening of potential biocides active in relation to corrosive biofilms and the determination of minimal concentrations of tested compounds provoking the death of corrosive biofilms were demonstrated in chapter four. The investigation of pilot systems modeling the field trials as well as the study of kinetics of corrosive biofilm formation in the flow systems were carried out applying bioluminescent method of intracellular ATP determination. The main factors accelerating the biofilm development were revealed.

Chapter five originally demonstrates the ability of two microbial enzymes – phospholipase D (PLD) from *Streptomyces netropsis* and nucleoside phosphotransferase (NPT) from *Erwinia herbicola* to transform a broad range of β-L-nucleosides into corresponding 5'-monophosphates and 5'-phosphatidyl derivatives. Studies on substrate specificity of PLD and NPT have revealed certain regularities. PLD was able to carry out rather effective transfers of phosphatidyl residue from egg lecithin onto all tested β-L-nucleosides. The crucial role of the 2'-hydroxyl group for demonstration of substrate or inhibitory properties by β-L-nucleosides was originally established with respect to NPT

The use of poly(vinyl alcohol) cryogel as a carrier for cell immobilization and the improvement of physiological state of yeast cells prior to their immobilization enable the obtaining of stable biocatalyst with varied shape and size of granules as shown in chapter six. The physiological state of free and immobilized cells was controlled determining the intracellular ATP concentration by bioluminescent method. The duration of wine fermentation was 3÷5 weeks at 15÷20°C, when the used concentration of biocatalyst was 4÷20 g l wine. The CO_2 pressure accumulated in the bottles was over 500 kPa at the end of the fermentation. The concentration of free cells in the sparkling wine was less than 9×10^3 cells/ml wine. The developed biocatalytic system could successfully be used in the processes of sparkling wine production even after a few months of storage at –20°C.

In: Biocatalysis and Biocatalytic Technologies ISBN 1-60021-041-4
Editor: G. E. Zaikov, pp. 1-8

Chapter I

The Nanoparticles, Possessing Low Curie Temperature, as Means of Self-Controlled Inductive Heating of Tumours

***F. S. Bayburtskiy*[1,3*], *L. A. Goncharov*[1], *D. B. Korman*[1,2], *N. A. Brusentsov*[1,2], *O. A. Shlyakhtin*[3], *A. E. Chekanova*[3], *V. A Naletova*[3] and *V. A. Turkov*[3]**

[1]N. M. Emanuel Instituite of Biochemical Physics, Moscow, Russia
[2]N. N. Blokhin Russian Oncology Research Center, Moscow, Russia
[3]M. V. Lomonosov Moscow State University, Moscow, Russia

Abstract

The present article investigates nanoparticles, possessed by low Curie temperature, which can be used as self-controlled mediators in the treatment of tumours by a method of magnetic hyperthermia. Authors carried out the synthesis of nanoparticles, made the physical and chemical analysis, and

---

* E – mail: Chembio@sky.chph.ras.ru, Bayburt@mail.ru

were offered the device, allowing qualitatively and quantitatively to estimate efficiency of a method of magnetic hyperthermia.

Keywords: nanoparticles, magnetic fluids, saturation magnetization, tumour, Curie temperature, magnetic hyperthermia.

Introduction

Hyperthermia is a rapidly developing technique in cancer therapy. It takes advantage of the higher sensitivity of tumour tissue to heat and typically involves heating of the affected organ to 43 – 45°C. Magnetic fluid hyperthermia [1] attracts increasing attention, since it allows minimizing side effects by the localized heating of only desired parts of the organism, including tumours located deep inside the patient's body. The method involves introduction of ferromagnetic particles (mediators) into the desired part of the organism and heating them with an alternating electromagnetic field of radio-frequency range (RF). Magnetic fluids based on nanocrystalline Fe_3O_4, stabilized by biocompatible surfactants [2 – 6], are typically used as mediators. Unfortunately, it is impossible to control the local temperature near the mediator particles, possibly causing local overheating and necrosis of normal tissue. This problem could be solved with ferromagnetic particles of high RF absorption and a Curie temperature (T_C) 42 – 45 °C. Thus, local temperature control can be ensured even with a nonuniform distribution of mediator particles throughout the tissue, variable RF intensity and uneven dissipation of the evolving heat. If T_C of this material can be adapted to the particular medical application.

A similar approach is used in cancer thermal therapies, which utilize implantable «thermoseeds» with a particular Curie temperature [7,8]. These macroscopically sized solid units are surgically inserted in the affected organ and are heated by an external RF. The T_C of the «thermoseed» determines the temperature of the heating. Significant limitations of these methods are the necessity of a stressful surgical intervention and a relatively small volume around each «termoseed» which can be effectively heated. The reduction of size of the heat-producing elements so that they can be delivered as a suspension via catheter into the bloodstream of the tumour and spread throughout its capillaries, is the obvious solution of these problems. Sato et al. [9,10] reported using 50 μm flakes of an amorphous ferromagnetic metal alloy with T_C 45°C for intratissue hyperthermia in dogs. However, these relatively large electrically conducting particles are heated in an RF field by eddy currents even above T_C and do not penetrate into small capillaries. Numerous studies [1 – 6, 12, 13] have shown that submicron-sized non-conducting particles are most effective for hyperthermia. The goal of this work was to produce biologically compatible, electrically non-conductive nanoparticles with T_C in this range do not exist, but by combining several elements and by varying the composition of the mixture it is possible to produce alloys, amorphous structures, ferrites and other multi-component system consisting of metals and metalloids. We used several synthesis technique to produce a variety of ultradisperse particles with suitable T_C and tested their behavior in an RF field.

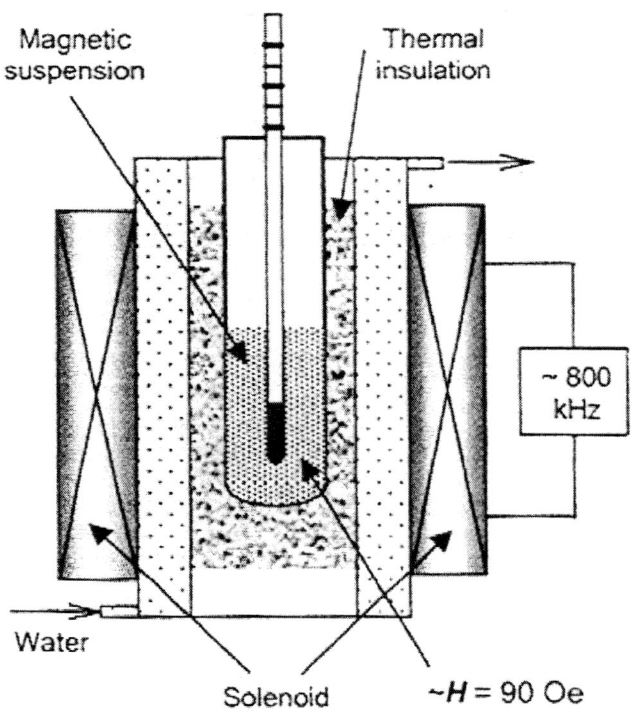

Figure 1. Scheme of the experimental setup.

Materials and Methods

In order to study their RF absorption rate, fine powders of $ZnFe_2O_4$ (T_C = 100 – 102°C), $La_{0.8}Sr_{0.2}MnO_3$ (T_C = 48°C) and $La_{0.75}Sr_{0.25}MnO_3$ (T_C = 56°C) have been prepared by the freeze-drying synthesis technique [11]. During absorption studies 0.5 g of each powder has been ultrasonically dispersed in 3 ml of water and placed in a glass test tube inside an air-cooled inductor (inner diameter 60 mm; length 200 mm) with a matching high-Q-resonator fed with RF power [12] (Figure 1). To avoid RF absorption by metal parts, the temperature of the fluid was monitored with an alcohol thermometer. A sample of magnetic fluid containing dextran-coated nanocrystalline Fe_3O_4 [4,12] was used for comparison. Previous studies have confirmed the absence of RF absorption in this frequency range by water or by components of the measuring cell [12].

Results and Discussions

Experimental heating curves are presented in Figure 2. Fe_3O_4 – based magnetic fluid demonstrates a fast, almost linear heating rate, close to that shown in [13], without any significant decrease in the tested experimental conditions. RF irradiation of the $ZnFe_2O_4$ suspension leads to significant, but less intense heating with no obvious temperature limit over the time of the experiment. This curve, taking into account the lower absorption rate of $ZnFe_2O_4$, is also similar to those of typical magnetic ferrofluids [13]. Both samples of La-Sr manganite powders demonstrate intense heating during the initial stages of the process, followed by temperature stabilization at 46.3°C and 37.8°C, respectively. These temperature correspond rather well with the T_C of the samples. The observed difference between T_C and the temperatures of RF absorption termination can be attributed to the sharp decrease of saturation magnetization M_S – usual for ferromagnetics in the vicinity of T_C and predicted by ferromagnetic exchange theory – and to the heat exchange balance. These results create many prospective applications of magnetic particles with predetermined T_C for heating of biological tissues and other objects to the optimum temperature using the demonstrated parametric feedback.

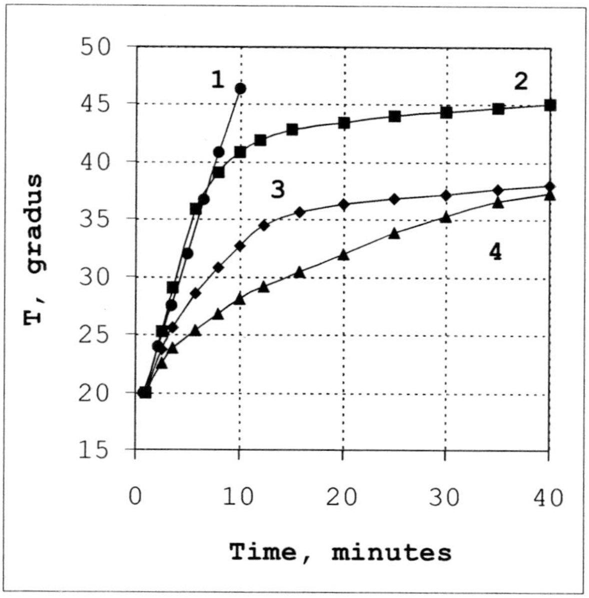

Figure 2. Time course of the temperature inside the measuring cell during RF (800 kHz) heating using the following mediators : 1 – Dextran-coated Fe_3O_4; 2 – $La_{0,75}Sr_{0,25}MnO_3$; 3 – $La_{0,8}Sr_{0,2}MnO_3$; 4 – $ZnFe_2O_4$.

Absorption of RF by ferromagnetic particles of various diameters is known to involve several physical mechanisms, including different types of remagnetization processes [5]. Usually RF absorption strongly depends on the crystallite size. Absorption rates of superparamagnetic nanoparticles are usually much higher than those of multi-domain crystallites, mostly due to Neel's losses. Powders produced by freeze-drying, are usually substantially agglomerated [14]. Intense milling of $La_{0.75}Sr_{0.25}MnO_3$ powders in a high-energy planetary ball mill resulted in a decrease of average aggregate size from 2 to 0.15 – 0.20 microns, as measured by light scattering. This is close to the average crystallite size observed by SEM (0.1 – 0.2 microns, Figure 3). Meanwhile, this treatment did not affect the RF absorption rate of this powder, with the heating curve being practically the same as that of the initial sample (data not shown). Since the observed size of crystallites even after milling is still much larger than the typical size of superparamagnetic particles, the most probable dominating absorption mechanism for these powders is scattering by displacement of magnetic domain walls. However, RF absorption rates of polycrystalline manganites are anomalously high compared to those of Fe_3O_4 polycrystals.

From our results we can conclude, that powders of La-Sr manganites make it possible to heat a sample by RF to the necessary temperature without exceeding it. Further heading will be automatically swished off a T_{LIM} , which is closely related to TC. Tc of La-Sr manganites strongly depends on the content of Me in $La_{1-X}Me_XMnO_3$ (where Me = Sr, Ba, Pb, Ag, Na). By varying the composition of the manganite it is possible to create materials with T_C ranging from 20 to 90°C [15]. Thus , RF heating to any desired temperature in this range without overheating can be achieved without any external temperature control. Application of $ZnFe_2O_4$ for these purpose is less attractive due to severe dependence of T_C on the metastable, poorly reproducible cation disrtibution between sub-lattices of the spinel structure, and a lower absorption rate, while no advantages over existing Fe_3O_4-based ferrofluids were observed.

Figure 3. SEM micrograph of $La_{0.75}Sr_{0.25}MnO_3$ powder after milling.

The prospects of a direct manganites application in RF hyperthermia of tumours will depend on the results of the ongoing medical compatibility studies.

Conclusion

We have proven that the freeze-drying synthesis technique permits the production of ultradisperse particles with the predetermined composition and T_C. The particles have a high absorption rate at lower temperature and abruptly stop absorbing RF energy at a particular temperature, preventing further heating of the sample. We continue searching for more biologically compatible nanoparticles for self-regulating hyperthermia.

The proposed RF heating with parametric feedback can be useful not only in medicine, but also in solving technological problems, for precise localized temperature control chemical and biochemical reactors. Coating of finely dispersed magnetic particles with a catalyst (including enzymes) combined with RF heating could ensure constant temperature at the reaction zone even in intense mass flows and strongly varying heat supply.

References

[1] Jordan, R. Scholz, R. Felix. // *J. Magn. Magn. Mat.* 201 (1999) 413 – 419.

[2] R. K. Gilchrist. // *Ann. Surgery*. 146 (1957) 596 – 606.

[3] M. Shinkai. // *J. Magn. Magn. Mat.* 194 (1999) 176 – 184.

[4] F. S. Bayburtskiy and N. A. Brusentsov. // *J. Magn. Magn. Mat.* 194 (1999) 83 – 89.

[5] R. Hergt. // *IEEE Trans. Magnetics*. 34 (1998) 3745 – 3754.

[6] D. C. F. Chan, D. B. Kirpotin, P. A. // *Bull. in Sci. Clin. Appl. Magn. Carriers*, Proc 1st Int. Conf., Plenum, New-York, 1997, P.607 – 618.

[7] J. A. Paulus. // *J. Endourology* 11 (1997) 295 – 300.

[8] T. C. Cetas, E. J. Gross, Y. Contractor. // *IEEE Trans. Biomed. Engineering* 45 (1998) 68 – 77.

[9] T. Sato. // *IEEE Trans. Magnetics*. 29 (1993) 3325 – 3330.

[10] H. Matsuki. // *Mater. Sci. and Engineering A*. 182 (1994) 1366 – 1368.

[11] O. A. Shlyakhtin, Y. J. Oh, Yu. D. Tretyakov. // *J. Eur. Ceram. Soc.* 20 (2000) 2047 – 2054.

[12] N. A. Brusentsov and F. S. Bayburtskiy. // *J. Magn. Magn. Mat.* 225 (2001) 113 – 117.

[13] R. Hiergeist. // *J. Magn. Magn. Mat.* 201 (1999) 420 – 422.

[14] Yu. D. Tretyakov, N. N. Oleynikov, O. A. Shlyakhtin. *Cryochemical Technology of Advanced Materials*, 1997, Chapman Hall, London.

[15] N. Rao, R. Mahesh, A. K. Raychaudhuri, R. Mahendiran. // *J. Phys.Chem. Solids*. 59 (1998) 487 – 502.

In: Biocatalysis and Biocatalytic Technologies ISBN 1-60021-041-4
Editor: G. E. Zaikov, pp. 9-21

Chapter II

Phenylalanine – Ammonia-Lyase of *Rhodotorula Aurantiaca*: Purification and Primary Characteristics

I. L. Bazukyan [1*]***, A. A. Hambardzumyan*** [2] ***and A. E. Vardanyan*** [1]

[1]Yerevan State University,Yerevan, Armenia
[2]SRI "Biotechnology", Yerevan, Armenia

Abstract

The phenylalanine – ammonia-lyase (PAL) of *Rhodotorula aurantiaca* KM-1 was isolated and purified up to homogeneity. The molecular weight of this tetramer enzyme was 291 kDa, and it comprised homologue subunits with molecular weight of 80 kDa. The isoelectric point of the enzyme was reached at pH 5,3. Polyclonal rabbit antibodies to PAL were obtained. The enzyme expressed maximal activity in reaction to de-amination under 50°C and pH 8,2. The enzyme deaminated L-phenylalanine and L-tyrosine. D-

---

* *E-mail:* physiol@ysu.am

phenylalanine, L-tryptophan and L-dioxyphenylalanine were not substrates for this enzyme.

Key words: phenylalanine - ammonia-lyase, *Rhodotorula aurantiaca*, subunit structure, polyclonal antibody.

Introduction

Phenylalanine – ammonia-lyase (EC 4.3.1.5) (PAL) catalyses deamination of L-phenylalanine to trans-cinnamic acid and ammonia. This enzyme is of high interest for clinicians in phenylketonuria treatment [1, 2] and for biotechnology during synthesis of L-phenylalanine from trans-cinnamic acid [3]. PAL was discovered in plants [4], yeasts [5], fungi [6] and actinomyces [7]. This is one of the rare enzymes which modify amino acids and do not contain pyridoxal 5'-phosphat. Instead, it contains an unusual prosthetic group – dehydroalanine [8].

The regulation of plant PAL was studied in details. The intricate mechanism of enzyme induction in response to many environmental stimuli in plants was defined [9-11].

There were well-known producers with high PAL activity [3, 12, 13]. The conditions of induction of enzyme were studied in detail both at the enzyme activity level [5, 14, 15] and at the transcription level [6, 16]. It was defined that L-phenylalanine was the inductor for this enzyme.

The PALs are from different sources, including yeasts, were purified and described in detail [17-19]. In particular, it was shown that the enzyme of *Rhodotorula glutinis* was homotetramer with a molecular weight of 330 kDa and the isoelectric point (p*I*) of this enzyme was reached at 5.4 [18]. As a result of the action of various proteinases on the PAL of *Rh. glutinis,* its peptide map was described [20].

A lot of works were dedicated to the study of the opportunity of using yeasts and/or fungous PAL for L-phenylalanine synthesis from cinnamic acid [21-25]. The optimal parameters for biotransformation were defined and the ways of stabilization of catalyst were identified.

The aim of this work is to isolate, to purify and to obtain the polyclonal rabbit antibodies to PAL as well as to study some physical and chemical and catalytic properties of PAL from *Rh. aurantiaca* KM-1 [12, 13].

Materials and Methods

Cell culture

Rh. aurantiaca KM-1 cells were grown and kept on slant beef extract agar (BE-agar). The subculturing was carried out twice a year. The suspension of the inoculum's cells grown on slant BE-agar was introduced then into 750 ml flasks with 100 ml of liquid nutrient medium for 22-24 hours. The inoculums were grown in tubes with slant BE-agar. The yeasts were cultivated on round rocking at rotation speed of 200 revos and 30^{0}C. The obtained biomass was separated by centrifugation at 10,000 g during 40 min. and at 4^{0}C, and stored in frozen condition at –18^{0}C. The different media like L-agar, L-broth, minimal agar (MA), minimal broth, Saburo's medium [26] were tested also both with adding the inductor – 2 g/l L-phenylalanine and without it.

All isolation and purification steps of PAL were realized at 4^{0}C. The cells disintegration was carried out by ultrasonic processing during 30 min. at 20 kHz tone frequency and 300 W power in buffer A, which contained 50 mM Tris-HCl pH 8.9, 1 mM ethylene diamine tetraacetate (EDTA), 0.2 mM phenylmetylsulphonylfluoride (PMSF). The cells debris were eliminated by centrifugation.

Enzyme Activity

The enzyme activity was measured by accumulation of cinnamic acid according to the modified method of Hyodo [27]. The calculation was based on a molar extinction coefficient for cinnamic acid equal to 10000 $cm^{-1}M^{-1}$ at 290 nm of wave length. The assay mixture contained Tris HCl, pH 8,9, 15 мM L-phenylalanine and necessary quantity of the enzyme. The synthesis of 1 μM cinnamic acid during 1 min. at 30^{0}C and pH 8,9 was accepted as a unit of enzyme activity.

Protein Content

The concentration of enzyme was measured by the methods of Groves and Davis (by the difference of absorption at 224 and 236 nm) or/and of Bradford (by the difference of absorption of kummasy brilliant blue and its complexes with proteins) [28]. The bovine serum albumin was used as a standard for enzyme concentration measuring.

Enzyme Purification

1. The method of ion-exchange chromatography was used in the first stage of purification. The rude enzyme extract was precipitated on the diethylaminoethyltoyopearl (DEAE-Toyopearl) 650 M column (2.5 x 15 cm) and equalized by buffer A. The elution was realized by 500 ml of linear gradient of NaCl (0 - 0.4 M), which was prepared on the same buffer. The active fractions (8-10 ml) were collected by automatic collector. They were combined and used in the next stage.
2. The second stage was adsorption on hydroxyapatite. The combined active fractions from the first stage of purification were precipitated on the column (2.5x7 cm). The enzyme was eluted by 200 ml of linear gradient of phosphate buffer pH 7.3 (0.01-0.3 M), which contained 1 mM EDTA and 0.2 mM PMSF. The fractions with PAL activity were combined and desalinated on the Sephadex G-25 column and equalized by buffer A.
3. The third stage was the treatment with the medical kaolin in batch process. For this purpose the desalinated preparation of PAL was mixed with dry kaolin (10 mg enzyme per 500 mg kaolin) and intermixed during 3-4 hours. After that the kaolin was removed by centrifugation. This preparation of PAL was concentrated on the DEAE-Toyopearl small column (1x5 cm).

Antibodies and their Analysis

The immunization of the rabbit was undertaken by modified method of Chen et al.[9].

At first, the mix of 1 ml of PAL (50 μg) and the complete Freund's adjuvant was injected into the rabbit. The injection was carried out subcutaneously in two points of the vertebral column. The injection was repeated 4 days later. The next injection was after a month with the same quantity of enzyme, but instead of complete Freund's adjuvant, 1ml of the incomplete one was used. The forth injection was repeated in the same way 4 days later.

The second stage of immunization was carried out by the intramuscular injection in two points of the hip, a month later after the last subcutaneous injection. Injection mixture was composed of 125 μg of enzyme and the incomplete Freund's adjuvant. The probe of blood was taken from ear-vein 4 days later, and this probe was used for the reaction of immunoprecipitation according to Uoxterlony [29].

The third stage of immunization was the intravenous injection of 125 μg PAL with physiological/salt solution 10 days later after the last one. It was carried out for the intensification of immune response and the synthesis of large quantity of antibodies.

About 20 ml of blood was taken from the heart of rabbit for purification of PAL's immunoglobulins G (Ig G). Ig G were purified by fractionating with $(NH_4)_2SO_4$ and by ion-exchange chromatography. The titer of antibody was fixed according to the method of Uoxterlony.

Molecular Weight

The molecular weight of the enzyme was determined by gel filtration. For this purpose, the enzyme preparation was passed through the 1,5x54 cm column packed with toyopearl 55 F. The column was calibrated by using the following protein standards with known molecular weight: ferritin (450 kDa), catalase (240 kDa), aldolase (60 kDa), cytochrome C (12.4 kDa).

Molecular Weight of Subunits

The molecular weights of the enzyme subunits were determined by sodium dodecyl sulfate polyacrilamide gel electrophoresis (SDS-PAGE, 12.6% polyacrilamide). The gel was calibrated with a mixture of a low molecular weight protein standard (Farmacia, Sweden): rabbit phosphorylase B (94 kDa), bovine serum albumin (BSA, 67 kDa), egg albumin (43 kDa), bovine erythrocyte carbonic anhydrase (30 kDa), soybean trypsin inhibitor (20.1 kDa), and bovine α-lactalbumin (14.4 kDa).

Isoelectric Point

The p*I* of the PAL was determined by vertical isoelectric focusing in plate of 5% polyacrilamide gel by the method of Farmacia firm. The gel contained 1.6% of ampholines LKB (Sweden), 75% of which had pH 4-6 and 25% - had pH 3.5-9. Some filter paper imbued with 0.1M NaOH and 0.1M H_2SO_4 was used as electrode buffers. And the mixture of wide p*I* spectrum proteins (Farmacia) was used as the pH markers.

Temperature and pH Optimums

For determination of temperature and pH optimums, the activity of enzyme was measured by the above-mentioned method at the range of temperature from 4 to 85^0C and the range of pH from 5 to 12. The reaction mixture for determination of the optimal pH contained 100 mM Tris-citrate buffer. Different values of pH were obtained by adding appropriate amounts of NaOH or HCl.

Substrate Specificity

The substrate specificity of PAL was determined by the measurement of enzyme activity in the presence of corresponding substrates.

Results and Discussion

Biomass

Determination of the PAL activity of *Rh. aurantiaca* cells, which were grown on the different media with L-phenylalanine and without it, showed that the adding of L-phenylalanine did not provoke the changes of the enzyme activity on full media. The activity of the enzyme increased on the synthetic media. While cultivating on BEB the PAL, activity of cells was the highest, so this medium was used for the biomass obtaining.

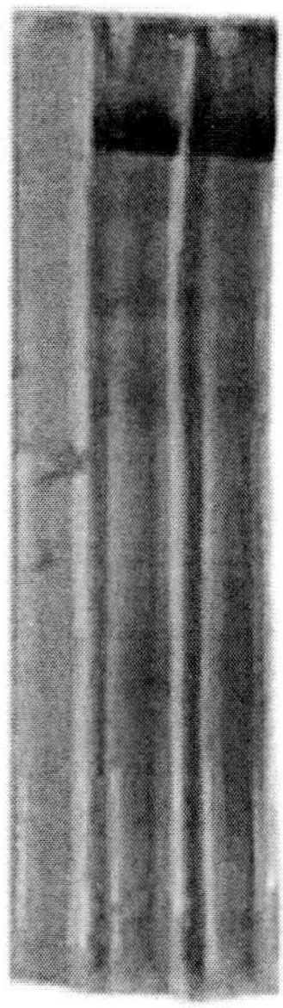

Figure 1. Disc-electrophoreses of PAL from *Rh. aurantiaca* KM-1 (1-control, 2-20 μg PAL, 3-60 μg PAL).

Isolation and Purification

The stages of the purification of PAL are presented in Table 1. Due to this procedure the specific PAL activity increased 36.6-fold and made 49.2% of total activity. This scheme of purification was rather effective in comparison with those ones described in literature. The purity of enzyme preparation was studied by disc-electrophoreses in plats of 7.7% PAAG with buffers system by Laemly (Figure 1). The picture showed that along with the line correlated with PAL Activity (proved in the separate experiment), there were several minor enzyme lines presented in gel. The homogeneity of the findings was estimated by us as 90%.

Table 1. Results of purification of PAL from *Rh. aurantiaca* KM-1

Steps	Enzyme, mg	Specific activity, U/mg	Income, %
Homogenate	739,1	0,08	100,0
Ion-exchange	99,3	0,51	85,6
Hydroxylapatite	20,0	1,75	59,0
Kaolin	11,5	2,93	56,8
Concentration	10,5	2,77	49,2

The Getting of Antibodies and their Analysis

The results of immunoprecipitation of purified antibodies in the horizontal agarous gel by method of Uoxterlony showed the purity and high titer (1:16) of polyclonal antibodies.

Molecular Weight

Figure 2 shows the results of determination of PAL molecular weight by method of gel-filtration. As it can be seen, the molecular weight of the PAL is equal to 291 kDa.

The Molecular Weight of Subunits

The molecular weight of PAL subunits is equal to 80 kDa (Figure 3). This data correlates with the literature ones. The sources describe the PAL as a tetramer molecule, which subunits molecular weight varies between 63-80 kDa, and the whole molecular weight of the enzyme is equal to 260-300 kDa. It should be noted that the molecular weight of subunits of active and inactive PAL preparations is different. As the results of SDS-PAAG electrophoresis (Figure 3) showed, the molecular weight of subunits of inactivated PAL preparation was equal to 68 kDa (6), whereas the subunits of active weigh of PAL preparation was equal to 80 kDa (7). Therefore, in this case the loss of activity of the protein was due to separation of polypeptide with weight equal to 12 kDa. It is the most likely, that this separation has happened under the influence of protease, based on which the following inhibitors of the protease - PMSF and EDTA - were used during PAL purification and concentration.

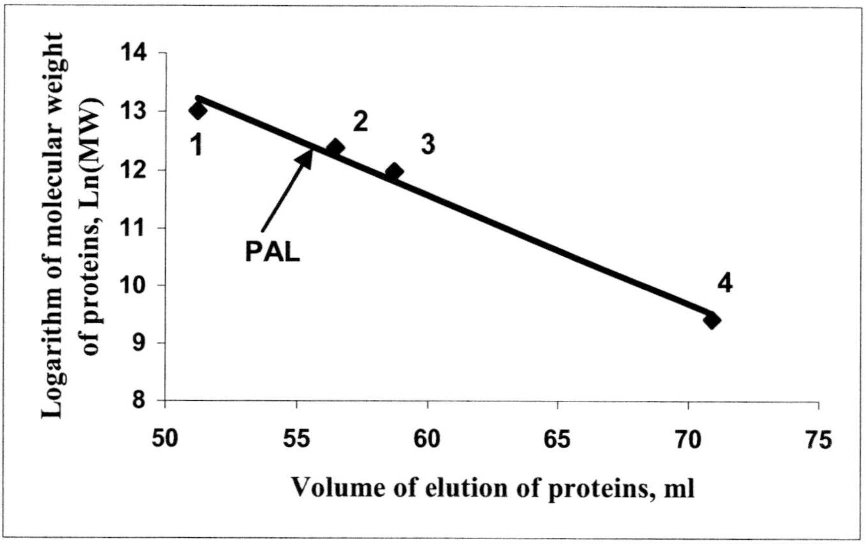

Figure 2. Determination of molecular weight of PAL from *Rh. aurantiaca* KM-1. 1- ferritin, 2- catalase, 3- aldolase, 4- cytochrome C, 5- PAL

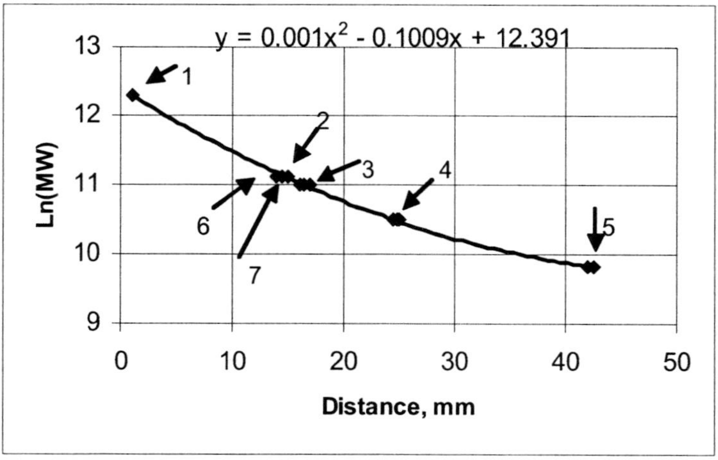

Figure 3. Determination of molecular weight of subunits by SDS PAAG-electrophores. 1 и 5 –ferritin, 2 –bovine albumine, 3 –catalase, 4 –lactate dehydrogenase, 6 –PAL, 7 –PAL-inactive.

The isoelectric point. The estimations show that the *pI* of the intact enzyme (PAL) from *Rh. aurantiaca* KM-1 is observed at pH 5.3 (Figure 4), which is in correspondence with the literature data for yeasts (*pI* is 5.4).

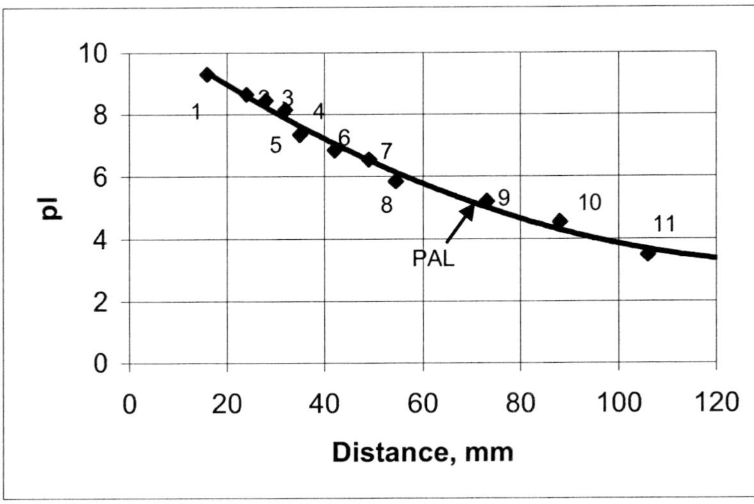

Figure 4. Isoelectric focusing of PAL. 1 - Trypsinogen (*pI*-9.30), 2 – lentil lectin basic group (*pI*-8.65), 3 – lentil lectin neutral group (*pI*-8.45), 4 – lentil lectin acid group (*pI*-8.15), 5 – myoglobin basic group (*pI*-7.35), 6 – myoglobin acid group (*pI*-6.85), 7 – human anhydrase (*pI*-6.55), 8 – bull anhydrase (*pI*-5.85), 9 - β-lactoglobuline A (*pI*-5.20), 10 – soybean trypsin inhibitor (*pI*-4.55), 11 - amyloglucosidase (*pI*-3.50)

The Specificity of Enzyme

According to the results presented in Table 2, the substitution of hydrogen by hydroxyl groups in aromatic ring of substrates of the row L-phenylalanine – L-tyrosine – L-dihydroxyphenylalanine (L-DOPA) decreased the PAL activity very distinctly. It seems that the aromatic ring of phenylalanine has an important role in the reaction catalyzed by PAL. These results correlate with the known data on substrate specificity of plant PAL. It is highly probable, that this property of PAL from *Rh. aurantiaca* KM-1 will be of great importance in the process of getting the L-tyrosine from suitable predecessor.

Table 2. Substrate specificity of PAL from *Rh. aurantiaca* KM-1.

Substrate, 5mM/l	Relative activity, %
L-phenylalanine	100,0
D- phenylalanine	0,4
L-tyrosine	53,1
L-tryptophan	0,0
L-DOFA	0,0

The Temperature and pH Optimums

The Figure 5 demonstrates the results of experiments on determination of temperature optimum of PAL. The PAL from *Rh. aurantiaca* KM-1 is the most active at 50-55^{0}C. The highest activity of enzyme was optimal at pH from 8.2 to 9.2 (see Figure 6). As it can be seen, the area of the highest activity of the enzyme is in somewhat alkaline zone. which is supported by literature data on yeast and plant PAL.

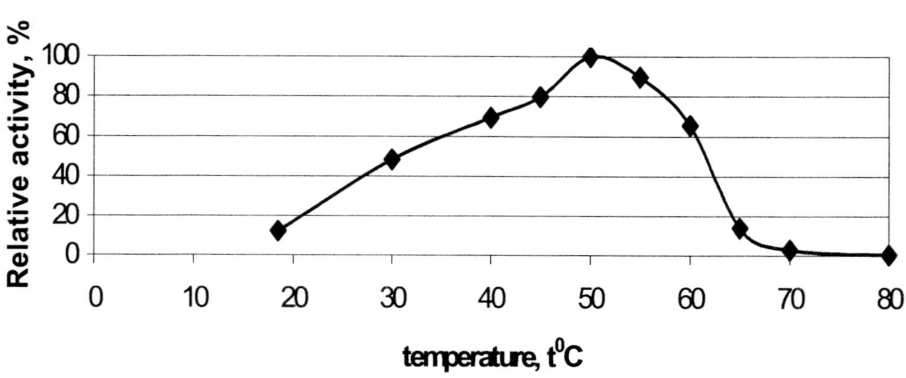

Figure 5. Temperature optimum of PAL from *Rh. aurantiaka* KM-1.

Therefore, the results of determination of properties and stabilization of PAL from *Rh. aurantiaca* will allow for expanding the application of this enzyme in biotechnology.

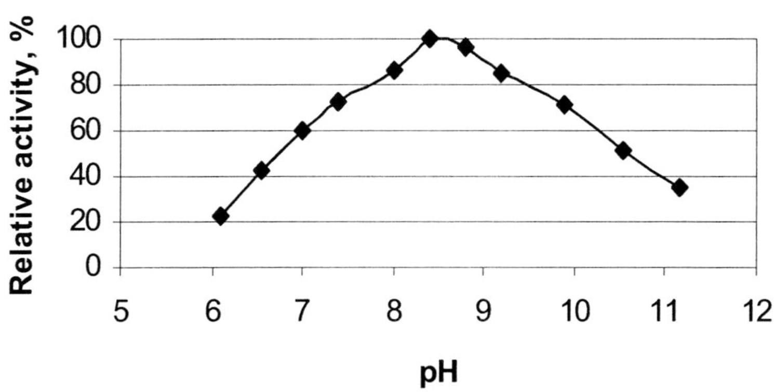

Figure 6. pH optimum of PAL from *Rh. aurantiaka* KM-1.

References

[1] H. L. Levy: *Proc. Natl. Acad. Sci. USA*, 96, 1811–1813 (1999).

[2] C. Sarkissian, Z. Shao, F. O. Blain, R. Peevers, H. Su, R. Heft, T. S. Chang, and C. R. Scriver: *Proc. Natl. Acad. Sci. USA*, 96, 2339–2344 (1999).

[3] C.T. Evans, K. Hanna, D. Conrad, W.Peterson and M. Misawa: *Applied Mikrobiol. Biotechnol.*, 25, 406-414 (1986).

[4] G. Bolwell, J. Bell, C. Cramer, W. Schuch, C. Lamb, R. Dixon: *Eur. J. Biochem.*, 149, 411-419 (1985).

[5] W. C.Marusich, R. A. Jensen, and L. O. Zamir: *Bacteriol. J.*, 146 (3), 1013-1019 (1981).

[6] L.A. Sikora and G.A. Marzluf: *Bacteriol. J.*, 150 (3), 1287-1291 (1982).

[7] A.V. Ames, and L.C.Vining: *Canadian J. Biochem.*, 48, 613-622 (1970).

[8] B. Schuster, and J. Retey: *Proc. Natl. Acad. Sci. USA*, 92, 8433-8437 (1995).

[9] R. Chen, T. Chang and M. Liu: *Agric. Biol. Chem.*, 52 (9), 2137-2142 (1988).

[10] E. Minami, Y. Ozeki, M. Matsuoka, N. Koizuka and Y. Tanaka: *Eur. J. Biochem.*, 185, 19-25 (1989).

[11] M. Bevan, D. Shufflebottom, K. Edwards, R. Jefferson and W. Schuch: *The EMBO J.*, 8 (7), 1899-1906 (1989).

[12] N. Yu. Mushy, M. B. Kupletskaya: *Applied Biochem. Microbiol.*, 17 (16), 107-112 (1981). (in Russian).
[13] M. B. Kupletskaya, G. A. Dolnikova: *Applied Biochem. Microbiol.*, 28 (1), 50-54 (1992). (in Russian).
[14] H. J. Gilbert and M. Tully: *Bacteriol. J.*, 150 (2), 498-505 (1982).
[15] J.F. Kane and M.J. Fiske: *Bacteriol. J.*, 161 (3), 963-966 (1985).
[16] Gilbert H. J., Stepheson J. R. and Tully M.: *Bacteriol. J.,* 153 (3), 1147-1154 (1983).
[17] D.S. Hodgins: *Biol. Chem. J.*, 246 (9), 2977-2985 (1971).
[18] R.R. Fritz, D.S. Hodgins and C.W. Abell: *Biol Chem. J.*, 251 (15), 4646-4650 (1976).
[19] R.I. Monge, M. Lara and A. Lopez-Munguia: *Biotechnology Techniques*, 9 (6), 423-428 (1995).
[20] H.J. Gilbert and G.W. Jack: *Biochem. J.*, 199, 715-723 (1981).
[21] S. Yamada, K. Nabe, N. Izuo, K. Nakamichi and I. Chibata: *Appl. Environment Mikrobiol.*, 42 (5), 773-778 (1981).
[22] G. M. Klehars, S. Peters: *United States Patent* 4,562,151, Dec 31, 1985. 6p.
[23] C.T. Evans, D. Conrad, K. Hanna, W. Peterson, C. Choma, and M. Misawa: *Appl. Mikrobiol. Biotechnol.*, 25, 399-405 (1987).
[24] N. Onishi, K. Yokozeki, Y. Hirose and K. Kubata: *Agric. Biol. Chem.*, 51 (1), 291-292 (1987).
[25] Tozalakyan P.V., Dyukova K.G. *Armenian patent* No 414, (1997). (in Armenian)
[26] N.S. Egorov: *MSU*, Moscow, (1983), 222 p. (in Russian)
[27] Hyodo H.: *Biochem. J.*, 79, 277-282 (1976).
[28] Peterson G.: *Meth. Enzymol.*, 91 (1), 95-119 (1983).
[29] D. Ketty.: Mir, Moscow, *I*, (1991), 287 p. (in Russian)

In: Biocatalysis and Biocatalytic Technologies ISBN 1-60021-041-4
Editor: G. E. Zaikov, pp. 23-37

Chapter III

Biocatalytic Systems Based upon the Immobilized Cells Degrading Parathion and *p*-Nitrophenol

E. N. Efremenko *
The M.V. Lomonosov Moscow State University, Moscow, Russia

Abstract

The construction of the immobilized biocatalytic system on the basis of artificial microbial consortium for parathion biodegradation was carried out. Some conditions of coexistence of various cultures in a joint system were investigated. The positive influence of several microorganisms (*E. coli* DH5α, *Pseudomonas fluorescens* B-473, *Pseudomonas fluorescens* B-495) on the *p*-NP degradation by *Pseudomonas putida* cells were revealed. All these cultures were introduced into the joint microbial system and the optimized biomass content of the consortium was established. The entrapment of the developed complex biocatalytic system into the poly(vinyl alcohol) cryogel allowed for the creation of immobilized consortium. This immobilized biocatalyst was used for the biodegradation of PT and *p*-NP

* e-mail: efremenko@enzyme.chem.msu.ru

under the batch conditions and obvious advantages compared to the immobilized single culture *Ps. putida*.

Keywords: parathion, *p*-nitrophenol, biodegradation, organophosphorus hydrolase, immobilized cells, artificial microbial consortium

Introduction

The organophosphorus compounds (OPC) are widely used as pesticides in the agriculture. The intensity of their application and the low solubility of OPC in water cause the notable accumulation of organophosphorus neurotoxins in the treated soils, ground and river waters, foods [1-4].

The literature review revealed, that OPC such as parathion (*O,O*-diethyl *O-p*-nitrophenyl phosphothioate, PT) could be degraded by microbial consortia present in the soil [5-8]. As a rule, the metabolic products of one culture are the substrates for another in such microbial community. The synergism among cells allows to realize the biological transformation of some compounds that can not be degraded by one individual microorganism. The natural consortia usually consist of a large number of various microbial species from diverse taxonomic groups. The attempt to isolate such microbial community from the nature often leads to the loss of some notable components of consortium and, as result of that, to the disruption of trophic chains inside of the system. Usually, it appears that such complex system is very unstable under any storage conditions and in the biotechnological processes. Sometimes, the existence of consortium isolated from the environment appears to be enough stable under artificial conditions, but the intensity of bioremediation processes is not satisfied to be scaled up. The attempts to accelerate the processes inside of microbial system usually result in displacement of consortium internal balance.

It is known, that PT can be involved into the metabolic microbial pathways as sole carbon source [7,8] or as a cometabolit (source of phosphor) [9,10]. In both cases, the rates of PT degradation are rather slow [8-11]. The *p*-nitrophenol (*p*-NP) often accumulates in the environment [12-14] as a result of PT hydrolysis catalyzed by organophosphorus hydrolase (OPH, EC 3.1.8.1) [15]. The microbial biodegradation of *p*-NP is an object of environmental attention since the letter is ubiquitous contaminant.

This work was devoted to the development of stable artificial microbial consortium based on the use of well-studied microorganisms participating in the PT degradation. The principal scheme of PT biodegradation by artificial consortium is performed in Figure 1.

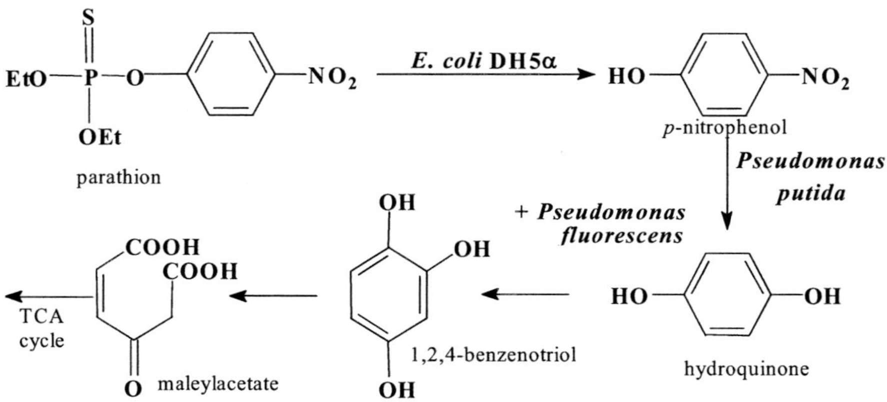

Figure 1. The proposal scheme of PT degradation by artificial microbial consortium.

According to the scheme, the two types of microbial cells were used for the PT biodegradation: the first type of bacteria should possess the OPH activity and hydrolyze the PT (recombinant strain *E. coli* DH5α [16]), whereas the second one should completely degrade the *p*-NP (*Pseudomonas putida*) [17]. Previously it was established, that this approach is effective enough [18,19].

The use of immobilized cells for biodegradation of various xenobiotics is the perspective approach to the environment decontamination [20]. The poly(vinyl alcohol)(PVA) cryogel is the most attractive matrix for immobilization of microbial cells, since the mechanically stable macroporous structure of this carrier allows to provide the process of PT biodestruction with good mass-transfer characteristics [21,22].

This work was aimed at the development of immobilized biocatalytic system on the basis of consortium immobilized in PVA cryogel decontaminating and the PT.

Materials and Methods

Bacterial Strains and Growth Conditions

The recombinant strain *Escherichia coli* DH5α producing the OPH was described previously [16]. The cells *Pseudomonas putida* B-1091, *Pseudomonas fluorescens* B-473 and *Pseudomonas fluorescens* B-495 were received from VKM (All-Russian Collection of Microorganisms).

All cultures of microorganisms were stored in TYI-agar-medium [23] at 4^0C. The same medium supplemented with 0.1-0.5 mM *p*-NP was used to grow the *Ps. putida* cells.

The overnight culture *E. coli* DH5α grown in the TYI medium without agar was inoculated (2% v/v) into 200 ml of Terr-medium [24]. The cells were cultivated on a shaker (220 rpm) at 32^0C up to the end of logarithmic growth phase, centrifuged (10 min, 6,000xg), washed with K-phosphate buffer (50 mM, pH 7.4) and used for immobilization or direct application in PT biodegradation. The same procedures were carried out with *Ps. fluorescens* B-473 and *Ps. fluorescens* B-495 cells, the accumulation of biomass was conducted in the TYI – medium.

The accumulation of biomass was monitored by the absorbency at 540 nm and the precise cell concentrations were calculated using calibration curves. To be used in *p*-NP degradation, the grown free cells were adapted at 32^0C for 25 h in MSM-medium [25] containing 0.3 mM *p*-NP.

Degradation of PT and p-NP

The PT degradation was spectrophotometrically monitored measuring the *p*-NP accumulation (405 nm). The probes with free cells were centrifuged (10 min, 6,000xg) before analysis. The absorbency of solutions with *p*-NP was measured in 20 mM CHES-buffer (pH 9.0, 17 000 $M^{-1}sm^{-1}$).

The MSM-medium prepared on the basis of phosphate buffer (20 mM, pH 8.0) [25] was used to investigate the *p*-NP degradation. Various *p*-NP concentrations were introduced into the MSM-medium sterilized antimicrobial Corning filters (0.20 μm, Corning Inc, Germany).

The initial velocities of the PT and *p*-NP degradation at various initial substrate concentrations were analyzed in the Lineweaver-Burk routine ($1/v_0$ vs. $1/[S]$). Every experimental point was the average of 3÷5 measurements.

Biocatalytic Systems Immobilization into PVA Cryogel

To obtain the immobilized consortium, the biomasses of tested cells were mixed in the various proportions in accordance with mathematical and statistical methods of experiment planning. Cell mixtures were thoroughly suspended in 10% aqueous PVA solution so as the total cell biomass in the granules was 10% (w/w). The immobilized biocatalytic systems on the basis of individual culture *Ps. putuda* and artificial consortium were obtained according to method described previously [23]. The diameter of obtained granules was 3÷4 mm.

The immobilized biocatalysts were used for *p*-NP degradation after adaptation to this substrate in the MSM-medium containing 0.1 mM *p*-NP (32^0C, 70 h).

The biocatalysts were washed with 20 mM phosphoric buffer (pH 7.4) after each batch-cycle.

Results and Discussion

The PT Hydrolysis Catalyzed by Cells with OPH Activity

According to preliminary data, the *E. coli* DH5α cells did not degrade the *p*-NP, but could effectively hydrolyze the PT. It was expected, that introduction of these cells into the artificial consortium could allow to realize the first step of whole scheme of PT degradation and to accelerate the slow process of *p*-NP mineralization. It was supposed, that these cells could consume together with *Ps. putida* cells the nonaromatic intermediates of *p*-NP degradation as carbon source and thereby the total balance in the system could shift towards the complete mineralization.

The analysis of literature data revealed the limited information concerning PT microbial hydrolysis. Since then, it was necessary to determine the velocity of PT hydrolysis catalyzed by cells with OPH activity and to compare it with the rate of *p*-NP degradation by *Ps. putida* cells in the proposed process.

The dependence of velocity of PT hydrolysis on the concentration of *E. coli* DH5α cells was studied. The culture was grown up to the end of logarithmic growth phase, which correlated with the maximal OPH cell activity. The PT concentration equal to 0.16 mM was chosen for this investigation, since that is maximal PT concentration deliquescent in the aqueous medium. It should be noted, that PT concentrations in the real objects usually are much less.

The kinetic parameters of PT hydrolysis were estimated in the presence of 25 mg/ml *E. coli* DH5α cells. The results (Figure 2) were obtained at pH 8.0, however the pH optimum of OPH activity is 9.0 [26]. The pH 8.0 was a compromise value for both parts of process, PT degradation and *p*-NP transformation, since *Ps. putida* cells accomplished the *p*-NP degradation only at pH 6.0-8.0 (data not shown). The analogue decision concerning the temperature conditions acceptable for all tested cultures was made and all experiments were carried out at 32°C.

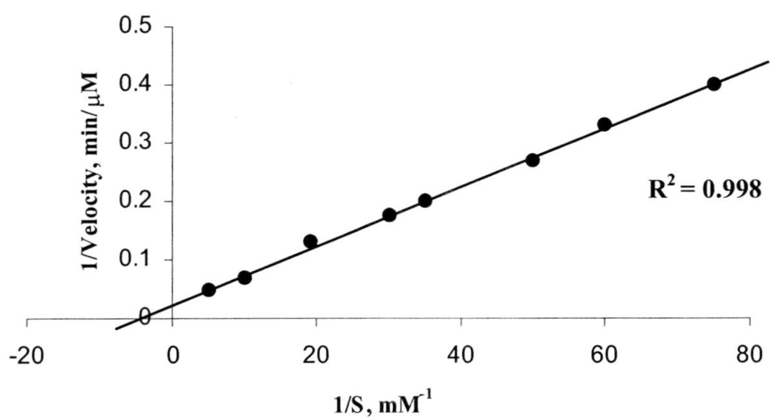

Figure 2. The parameters of PT hydrolysis catalyzed by *E. coli* DH5α cells possessing OPH activity.

It was established, that Km and Vmax of reaction of PT hydrolysis catalyzed by *E. coli* DH5α cells at pH 8.0 were 0.128 mM and 28 μM/min, respectively.

p-NP Degradation by *Ps. putida* Cells

The cultivation of *Ps. putida* cells was carried out in the MSM-medium with various *p*-NP concentrations (Figure 3). The growth of this culture in the medium containing *p*-NP was characterized by the long enough lag-period, and its duration correlated with the *p*-NP concentration. The duration of lag-phase was 2 days and a week, when the *p*-NP concentration was 0.75 mM (Figure 3) and 0.95 mM (data not shown), respectively. There was no cell growth in the medium with 1.2 mM *p*-NP.

The amount of cell biomass accumulated in the cultural media also depended on the initial *p*-NP concentration and varied from 0.3 up to 0.45 mg/ml. These results completely concurred with the literature data [7,10,25]. Small enough amount of accumulated cell biomass was shown to degrade the 80% of *p*-NP initially introduced to the media for 24-25 h of cultivation.

It was established that *p*-NP degradation could take place in the absence of cell growth. The addition of new portions of *p*-NP to the cultural media with grown cells resulted in the rapid substrate degradation. In this case the long lag-period of cell adaptation to the *p*-NP was absent (Figure 3). The data are performed for cultivation of cells in the medium with 0.1 mM, since the same tendency was observed for all tested *p*-NP concentrations. Since then, the adapted to *p*-NP and concentrated *Ps. putida* cells were used in the next experiments.

It was revealed [25], that concentration of inorganic phosphates is important for metabolic activity of cells degrading the *p*-NP. The influence of phosphate concentration on the rate of *p*-NP degradation was studied herein. The maximal rate of substrate oxidation by cells was established at 20 mM concentration of phosphate ions (data not shown). This concentration was used in all further experiments.

It was supposed that *p*-NP oxidation could be considerably intensified by the usage of high concentrations of *Ps. putida* cells, since the dependence of velocity of 0.5 mM *p*- NP degradation on the concentration of cell biomass was investigated (Figure 4a). The enhancement of cell concentration over 10 mg/ml appeared to be not expedient because it does not give notable increase in degradation velocity.

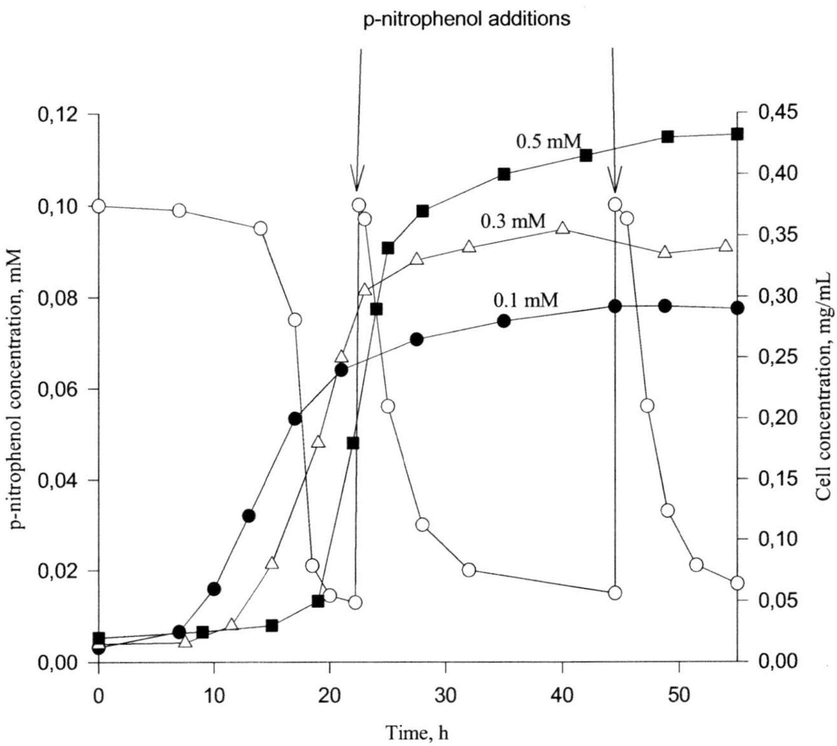

Figure 3. The growth of *Ps. putida* cells in the MSM-medium with various *p*-NP concentrations: 0.1 mM (●), 0,3 mM (△) and 0.5 mM (■), respectively; symbol O marks the degradation of 0.1 mM *p*-NP.

The use of cells (10 mg/ml) adapted to *p*-NP for its degradation showed that concentration of *p*-NP equal to 0.5 mM appeared to be optimal and guaranteed the maximal velocity 0.43 μM/min (Figure 4b). Comparing the velocities of PT hydrolysis and *p*-NP degradation it should be noted that the later process is obviously limiting in the functioning of the offered microbial system. Thereupon, the next steps of investigation were directed at the acceleration of *p*-NP degradation to increase the velocity of whole process.

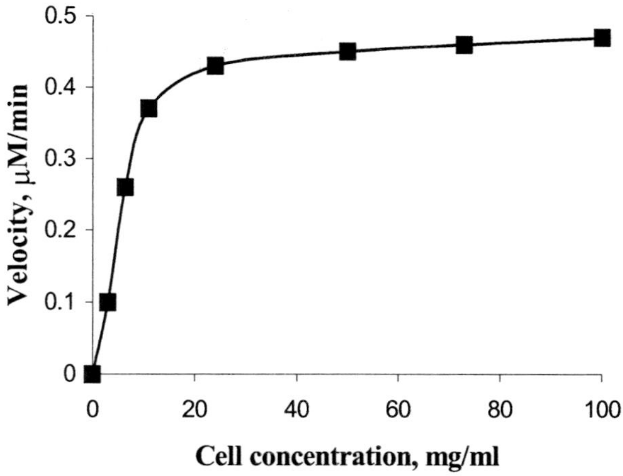

a

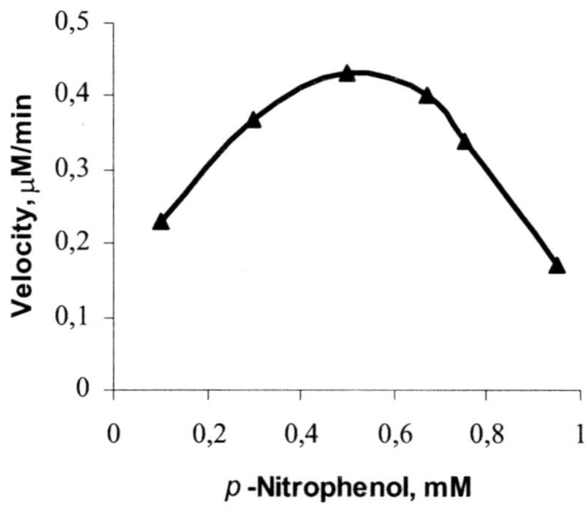

b

Figure 4. The dependence of velocity of *p*-NP degradation on the concentration of *Ps. putida* cells (*p*-NP concentration - 0.5 mM)(a) and the substrate concentration (cell concentration - 10 mg/ml) (b).

Interaction of Various Cultures with *Ps. putida* Cells in the p-NP Degradation

The presence of *E. coli* DH5α cells in the medium with *Ps. putida* provided a positive influence on the process of *p*-NP degradation (Figure 5). The best results were obtained, when two cultures, *E. coli* DH5α and *Ps. putida*, were added in equal concentrations to the cultural medium. Exactly these data defined the final choice of culture with PTE-activity for artificial consortium.

Usually, the natural consortia degrading the *p*-NP comprise several strains of genus *Pseudomonas* entering into close synergism [12]. The preliminary screening of *Pseudomonas* strains revealed two cultures which addition to the medium with *Ps. putida* cells accelerated the *p*-NP degradation (Figure 5).

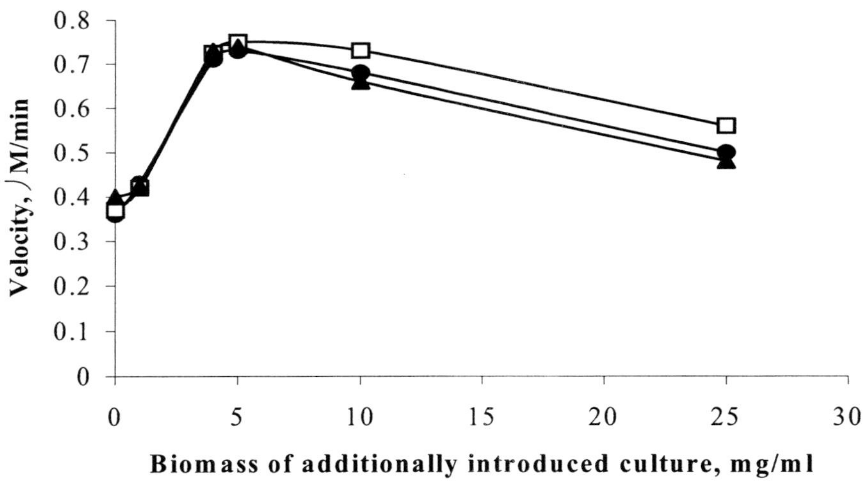

Figure 5. The influence of additionally introduced microbial cultures on the velocity of *p*-NP degradation by *Ps. putida* cells (The concentration of *Ps. putida* cells was 10 mg/ml). Variants of added cells were following: *E. coli* DH5α (▲), *Ps. fluorescens* B-473 (●), *Ps. fluorescens* B-495 (□).

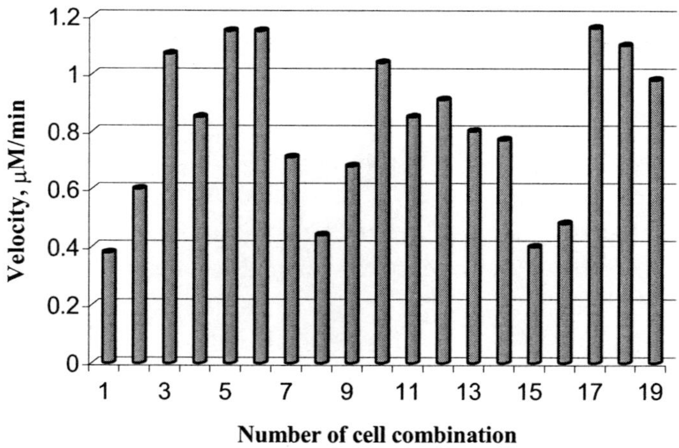

Figure 6. The velocity of *p*-NP degradation in the presence of additionally introduced various microbial combinations. Concentration of *Ps. putida* cells in each variant was constant and equal to 10 mg/ml. The mass content of each culture in the tested combinations were following (mg/ml): control (only cells *Ps. putida*)(1); a+b+c=5+5+0 (2); a+b+c=5+5+5 (3); a+b+c=5+10+5 (4); a+b+c=5+5+10 (5); a+b+c=5+10+10 (6); a+b+c=5+0+0 (7); a+b+c=5+10+0 (8); a+b+c=10+10+0 (9); a+b+c=10+10+10 (10); a+b+c=10+5+5 (11); a+b+c=10+5+10 (12); a+b+c=10+10+5 (13); a+b+c=10+5+0 (14); a+b+c=10+0+5 (15); a+b+c=10+0+10 (16); a+b+c=0+5+5 (17); a+b+c=0+10+5 (18); a+b+c=0+5+10 (19), where *E.coli* DH5α (a), *Ps. fluorescens* B-473 (b) and *Ps. fluorescens* B-495 (c).

The further study was aimed at the combination of all these cultures in one microbial system. It was necessary to optimize the quantitative ratios among them. The routine mathematical methods of experimental data treatment were used to vary the cell concentrations in the joint biocatalytic system. In all studied variants of combinations the cells *Ps. putida* were present as main culture in constant concentration (10 mg/ml) (Figure 6).

The mathematical treatment of obtained results gave the adequate equation of regression, which could be used as interpolation formula:

$$y = 9.806 - 0.739a - 0.181b + 0.814c + 0.334ab + 0.461bc,$$

where "y" is the velocity of *p*-NP degradation (μM/min), whereas "a, b, c" are the concentrations of cells, respectively, *E . coli* DH5α, *Ps. fluorescens* B-473 and *Ps. fluorescens* B-495 (mg/ml).

This equation reflects the role of individual culture in the developed complex biocatalytic system and the interactions between the certain strains inside it. The analysis of this mathematical description of obtained system proved that all tested cultures should be introduced into consortium. The optimal content of developed consortium was revealed and was following (mg/ml): *Ps. putida* cells - 10; *E . coli* DH5α - 5, *Ps. fluorescens* B-473- 5 and *Ps. fluorescens* B-495 – 10. So, the total mass of cells in this complex system had to be 30 mg/ml. Only its immobilization could allow the easy work with this complex cell mixture.

Immobilized Biocatalytic System on the Basis of Several Cultures

The immobilized biocatalyst on the basis of developed cell mixture was prepared concurrently with the biocatalyst containing only *Ps. putida* cells.

The granules of both biocatalysts in equal quantities were used to degrade the 0.1 mM *p*-NP. The conditions of experiments were the same as earlier for the free cells. Several consecutive cycles of batch incubation of the same probes of biocatalysts were conducted. The advantages of immobilized biocatalytic system in the process of *p*-NP degradation were obvious (Figure 7), although the rates of degradation increased from cycle to cycle for both biocatalysts.

The both immobilized biocatalytic systems were used to degrade the different *p*-NP concentrations. The immobilized consortium demonstrated higher velocities of *p*-NP degradation than individual immobilized culture (Table 1). The advantages of immobilized artificial consortium became evident with the increase of *p*-NP concentration in the medium. The tendency typical for the mixture of free cells was also observed for the immobilized consortium.

The investigation of PT degradation by the immobilized consortium revealed the 5-fold decrease in OPH activity of complex system compared to the free *E. coli* DH5α cells (data not shown). Probably, the high cell density inside of granules generated the diffusion difficulties and PT became less available for cells with OPH activity. Ultimately, this decrease in PT hydrolysis velocity had no serious influence on the total process, because the first step of PT transformation remained much quicker than the *p*-NP oxidation even with enhanced velocity.

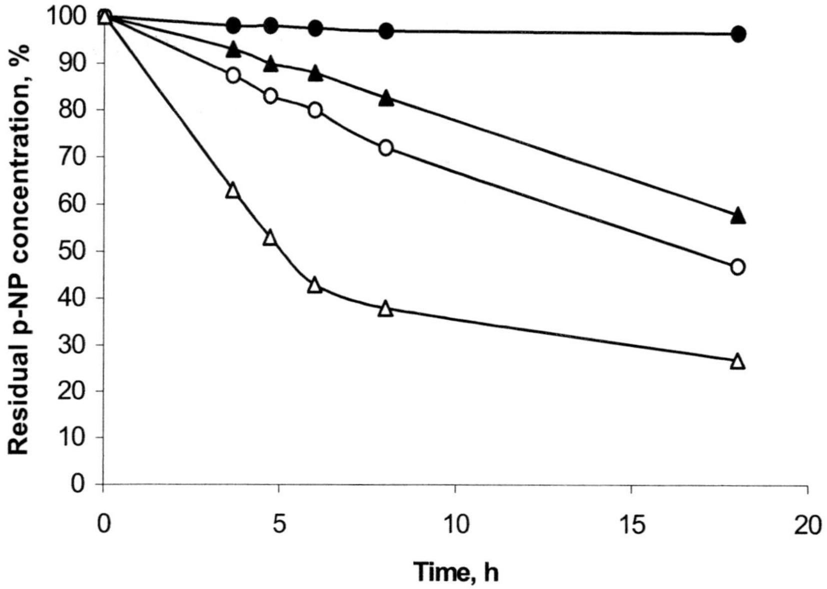

Figure 7. The degradation of 0.1 mM *p*-NP by the immobilized biocatalytic systems (500 mg/ml) under batch-conditions in the 1st and 3rd working cycles: immobilized individual culture *Ps. putida* (● and ○) and immobilized consortium (▲ and △), respectively.

Table 1. The influence of *p*-NP concentration on the velocity of its degradation under the action of immobilized biocatalytic systems on the basis of individual culture *Ps. putida* and artificial consortium.

Biocatalyst and its concentration	Concentration of *p*-NP		
	0.1 mM	0.3 mM	0.5 mM
	Velocity of *p*-NP degradation, mM/min		
Immobilized *Ps. putida* cells (500mg/ml)	$5.4*10^{-5}$	$9.7*10^{-5}$	$4.6*10^{-5}$
Immobilized consortium (500 mg/ml)	$6.4*10^{-5}$	$14.6*10^{-5}$	$7.8*10^{-5}$
Immobilized consortium (750 mg/ml)	$7.5*10^{-5}$	$20.7*10^{-5}$	$11.0*10^{-5}$

This immobilized consortium kept ability to degrade the PT and *p*-NP under batch conditions 6 months at the minimum. The growth of free cells or contaminants in the medium with PT and *p*-NP was not revealed during the all period of immobilized consortium use. That fact allowed to conduct the experiments with immobilized biocatalytic system under aseptic conditions.

Conclusion

It had to be noted that process of *p*-NP oxidation in the presence of immobilized consortium run with lower rates than that with free cells. Nevertheless, the possibility to stable exploitation of this immobilized biocatalytic system for a long time is important for its further application. The optimization of temperature conditions of cell immobilization as well as the size of granules should improve the mass transfer characteristics of immobilized biocatalyst and its biodegradation efficiency. This investigation can be regarded as a new approach to the elaboration of bioremediation technology.

References

[1] F.J Schenck, R. Wagner: *Food. Addit. Contam.*, 12 (4), 535 (1995)
[2] V. Leoni, A.M. Cariechia, R. Comi, F. Martini: *Bull. Environ. Contam. Toxicol.*, 54 (6), 870 (1995)
[3] A.J. Ali: *Anal. Lab.* (Bulg.), 2 (3), 210 (1993)
[4] C.P. Cai, M. Liang, R.R. Wen: *Chromatogrphia*, 40 (7/8), 417 (1995)
[5] K. Siddaramappa, P. Rajaram, N. Sethunathan: *Appl. Microbiol.*, 26, 846 (1973)
[6] K.L. Mick, P.A. Dahm: *J. Econ. Entomol.*, 63, 115 (1970)
[7] D.M. Munnecke, D.P.H. Hsieh: *Appl. Microbiol,* 28 (2), 212 (1974)
[8] L.M. Nelson: *Soil Biol. Biochem.*, 14 (3), 219 (1982)
[9] L.M. Nelson, B. Yaron, P.H. Nye: *Soil Biol. Biochem.*, 14 (3), 223 (1982)
[10] N.L. Rani, D. Lalithakumari: *Can. J. Microb.*, 40, 1000 (1994)
[11] C.G. Daughton, D.P.H. Hsies: *Appl. Environ. Microb.*, 34, 175 (1977)
[12] Rozenberg, M. Alexander: *Appl. Environ. Microb.*, 37, 886 (1979)
[13] M.A. Heitkamp, V. Camel, T.J. Reuterm, W.J. ADAMS: *Appl. Envir. Microb.*, 56 (10), 2967 (1990)
[14] J.C. Spain, D.T. Gibson: *Appl. Envir. Microb.*, 57 (1) ,812 (1991)
[15] W. Mulbry, P. del Valle, J. Karns: *J. Pest. Sci.*, 46, 35 (1996)

[16] S.D. Varfolomeyev, E.N. Efremenko, Y.A. Votchitseva, T.K. Aliev: *RU Patent* No.2232807 (2004).

[17] L. Harper, C.S. Mcdaniel, C. Miller, J. Wild: *Appl. Environ. Microb.*, 54 (10), 2586 (1988)

[18] A.W. Walker, J.D. Keasling: *Biotechnol.Bioeng.*, 78 (7), 715 (2002)

[19] E.S.Gilbert, A.W. Walker, J.D. Keasling: *Appl. Microbiol. Biotech.*, 61, 77 (2003)

[20] M.B. Cassidy, H. Lee, J.T. Trevors: *J. Ind. Microbiol.*, 16, 79 (1996)

[21] V.I. Lozinsky, F.M. Plieva: *Enzyme Microb. Tech.* 23, 227 (1998)

[22] V.I. Lozinsky, F.M. Plieva, A.L. Zubov: *Biotekhnologiya* (in Russian), 1-2, 32 (1995)

[23] E.I. Rainina, E.N. Efremenco, S.D. Varfolomeyev, A.L. Simonyan, J.R. Wild: *Biosens Bioelectron*, 11 (10), 991 (1996)

[24] A.L. Simonyan, E.I. Rainina, V.I. Lozinsky, I.E. Badalian, G.E. Khachatrian, S.SH. Tatikian, T.A. Makhlis, S.D. Varfolomeyev: Appl. Biochem. *Biotechnol.*, 36, 199 (1992)

[25] S.F. Nishino, J.C. Spain: *Environ. Sci. Technol.*, 3, 489 (1993)

[26] S.D.Varfolomeyev, T.K. Aliev, E.N. Efremenko: *Pure Appl. Chem.*, 76 (9), 1781 (2004).

In: Biocatalysis and Biocatalytic Technologies ISBN 1-60021-041-4
Editor: G. E. Zaikov, pp. 39-58

Chapter IV

Analysis of Bacteria Self-Immobilized into Corrosive Biofilms

E. N. Efremenko[a*], R. E. Azizov[a], V. M. Abbasov[b] and S. D. Varfolomeyev[a]

[a]The M. V. Lomonosov Moscow State University, Moscow, Russia
[b]Azerbaijan NAS, the Y. H. Mamedaliyev Institute of Petrochemical Processes,
Baku, Azerbaijan

Abstract

To determine the total concentration of cells composing the corrosive biofilms, a new approach was developed using bioluminescent method of ATP detection. The express screening of potential biocides active in relation to corrosive biofilms and the determination of minimal concentrations of tested compounds provoking the death of corrosive biofilms were demonstrated herein. The investigation of pilot systems modeling the field trials as well as the study of kinetics of corrosive biofilm formation in the

* efremenko@enzyme.chem.msu.ru

flow systems were carried out applying bioluminescent method of intracellular ATP determination. The main factors accelerating the biofilm development were revealed.

Key words: bioluminescent analysis, biocorrosion, biofilms, sulphate-reducing bacteria, intracellular ATP.

Introduction

In the environment, the microbial communities usually comprise mixed populations interacting with each other. Corrosive biofilms are among such microbial natural associations [1, 2]. Microbial corrosion is the result of numerous interactions between various biofilm components, which are in dynamic flux. Sulphate-reducing bacteria (SRB) are usually present in the corrosive biofilms on the surfaces of ship hulls, wastewater pipelines and oil industry equipment [3]. The SRB usually form the complex biofilms with heterotrophic cells [4] producing the organic acids necessary as substrates for SRB. The high concentrations of SRB provoke the quick development of metal corrosion. Thereupon, the quantification of SRB and other corrosive cells in sediments, wastewaters and fouling biofilms is of great practical and scientific relevance. Common microbial techniques based on selective cell cultivation are of limited usefulness for quantification and characterization of environmental populations since the used methods are time-consuming and inconvenient for field trials or express analysis. The methods based on the application of bioluminescent method of intracellular ATP determination could be used as alternative approach to the cell enumeration [5, 6].

This work was devoted to the development of a new express method for the investigation of corrosive biofilms using bioluminescent determination of intracellular ATP concentration. The main regularities of corrosive biofilm formation and the influence of different factors on the biofilm growth were studied herein. The determination of minimal inhibitory concentrations (MIC) of corrosion inhibitors possessing biocide activity in relation to biofilms was carried out.

Materials and Methods

Bacterial Cultures and Media

Desulfovibrio vulgaris B-1388 cells and *Rhodococcus ruber* AC-1513 cells were used in this work as representatives of SRB and heterotrophic microorganisms, respectively, widely spread in the objects subjected to biocorrosion. Both strains were taken from the VKM (All-Russian Collection of Microorganisms).

To grow the inoculum of *D. vulgaris* cells, the API medium was used (g/l): lactate – 4.0; yeast extract – 1.0; ascorbic acid – 0.1; $MgSO_4$ x $7H_2O$ – 0.2; K_2HPO_4 – 0.01; $Fe(SO_4)_2(NH_4)_2$ x $6H_2O$ – 0.2; NaCl – 10.0; pH 6.8-7.0. Cultivation of cells was carried out in the Hottinger's test tubes. The hot medium (17 ml) was saturated with argon to remove the soluble oxygen. After inoculation the *D. vulgaris* cells were grown at 30^0C under anaerobic conditions.

The *R. ruber* AC-1513 cells oxidizing the oil hydrocarbons were grown in the following medium (g/l): glucose - 50.0, sprat tryptic hydrolysate - 5.0; NaCl – 7.5; KH_2PO_4 - 3.0; Na_2CO_3 – 0.1; K_2HPO_4 – 6.0; NH_4Cl – 2.0; $MgSO_4$ x $7H_2O$ – 0.2; $CaCl_2$ x $6H_2O$ – 0.01; $MnSO_4$ x $5H_2O$ – 0.02; $FeSO_4$ x $7H_2O$ – 0.01; pH 6.7 – 7.2. Cultivation of hydrocarbon-oxidizing bacteria was undertaken in the 500-ml Erlenmeyer flasks on a shaker (Lab-therm, Adolf Kühner, Switzerland) under aerobic conditions with constant agitation (220 rpm) at 28 ^{0}C.

Natural association, isolated from the wastewaters of Almetevsk oilfield (Tatarstan, Russia) composed of SRB and heterotrophic cells was grown under the same conditions mentioned for *D. vulgaris* cells, but the medium was not boiled and saturated with argon. After inoculation, the cells were grown at 30^0C.

The biofilms were grown on the surface of steel coupons. According to the spark optical emission spectrometry, the composition of tested steel sample was following (w/w): Fe 98.48%, C 0.06%, Si 0.13%, P 0.03%, S 0.045%, Mn 0.51%, Cr 0.10%, Mo 0.02%, Ni 0.14%, and Cu 0.49%. The metal coupons were pretreated with 0.1 M HCl, washed with tap water and dried prior to use [7].

To grow the biofilms under the laboratory conditions the SRB and natural microbial association TAT were applied. The concentrations of different biocorrosive cells inside of formed biofilms were determined after 14 days of cell growth.

Intracellular ATP Determination

Dimethylsulfoxide (DMSO), sodium dodecyl sulphate (SDS) and cell releasing agent (CRA, New Horizons diagnostics Co., USA) were tested as ATP extracting agents (Figure 1). Each extracting agent (0.9 ml) was used to treat the 0.1 ml sample of cell suspensions. An aliquot (50 μl) of cell extract was added to the luciferase reagent (50 μl) and the intensity of bioluminescence was measured. The luciferase reagent developed at the Chemical Department of the M.V. Lomonosov MSU [8] and Bioluminometer 3550i (New Horizons diagnostics Co., USA) were used [5, 6].

The optimized conditions for ATP extraction from the biofilms were following (Figs.2, 3): the ratio of DMSO volume to square of biofilm surface was 0.5 ml/cm^2, the duration of ATP extraction was 10 min.

To quantify the cells in the biofilms and planktonic cultures the calibrations establishing the dependence of intracellular ATP concentration on cell concentration were used. The precise cell concentrations used for these calibrations were defined by the microbiological method [9].

The differentiative method of cell quantification (DMCQ) in the mixed cultures based on the complex approach, combining the microbiological assay with bioluminescent analysis and standard mathematical methods using to process the kinetic data obtained from bacteria growth curves was applied as published previously [10].

Flow Systems

Two flow systems were used in the work. The total volume of medium for SRB in the first system was 8.25 l, whereas the volume of working cell with metal coupons was 2.25 l. Sixteen metal coupons were used to investigate the kinetics of biofilm growth in the first system. The flow and dilution rates were 7.5 l/day and 3.3 day^{-1}, respectively. TAT with initial concentration of SRB equal to $7.0*10^3$ cell/ml was inoculated to the first system to grow the biofilm. The system was placed to the thermostatically controlled chamber at 28^{0}C. The concentrations of sulphide and cells accumulated in the biofilm content were controlled.

The second flow system used in the work was prepared with the same total volume (8.25 l), but the volume of working cell with 8 metallic coupons was notably less (100 ml). The flow and dilution rates here were 3.2 l/day and 35.5

day^{-1}, respectively. The main conditions of system functioning as well as the controlling parameters were the same as for the first system.

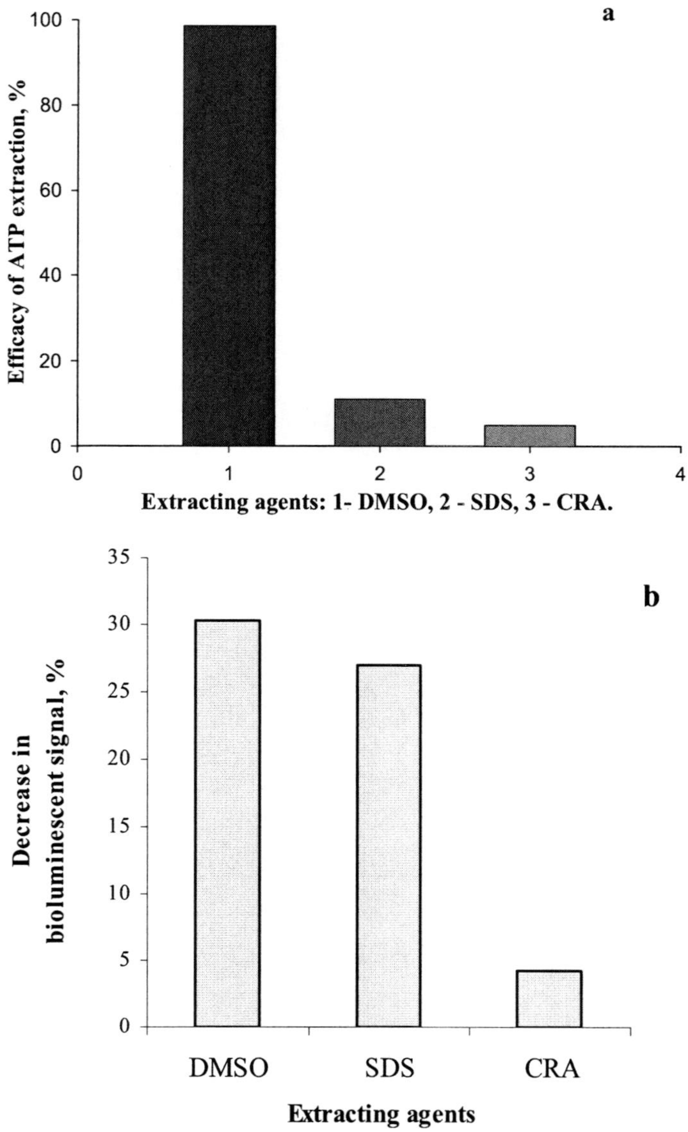

Figure 1. The influence of various extracting agents on the efficacy of ATP extraction from cells composing biofilms (a) and on the level of measured bioluminescence (b).

MIC Determination

To detect the MIC values, the biofilms were grown in the API medium for 2 weeks. The initial concentrations of cells in the biofilms taken for investigation were varied in the range of $(3.1\div5.3)*10^6$ cells/cm^2.

Since the SRB composed the TAT majority, the total cell concentration in the probes was estimated using the specific ATP concentration typical for this type of cells detected previously ($5.0*10^{-18}$ mole/cell) [5,6]. The following corrosion inhibitors (CI) were used in the work: Kathon (Interdisp, Russia); SXT1003 (Nalco, USA); Inhibitor 55 (IOPhC, Russia); Khazar, Nitro-1 (INKOR, Azerbaijan, [11]).

The biofilms in the presence of CI were exposed in the physiological solution (0.9% NaCl) at 28°C for 7 days (pH 6.0). The concentrations of tested CI were varied in a wide range (0-1200 ppm).

Sulphide Determination in Corrosive Biofilms

The sulphide determination was based on the detection of methylene blue formed as product of reaction between dimethylparaphenylene diaminesulphate (DMP) and iron (III) [10]. Metallic coupons (2.27 cm^2 in diameter) were treated with 5 ml zinc acetate solution and 4.95 ml DMP was added. Then 50 µl iron (III) ammonium sulphate was introduced to the mixture and the concentration of methylene blue was measured spectrophotometrically (Agilent 8453 UV-visible spectrometry system, Agilent Technologies, Germany, 660 nm) after 30 min.

Results and Discussion

Development of Bioluminescent Analysis for Biofilm Investigation

The experiments were aimed at the elaboration of method of total express quantification of microbial cells, present in the content of biofilms and provoking the biocorrosion. The influence of several factors on the effectiveness of ATP extraction from the cells localized in biofilms (Figure 1a) and on the level of measured bioluminescence was studied (Figure 1b). The nature of extracting agent (Figure 1), the ratio between the surface of treated coupon covered with biofilm (Figure 2) and the duration of biofilm treatment with extracting reagent (Figure 3) were among the investigated factors.

The ATP extraction from the biofilm cells was conducted applying DMSO, SDS and CRA. Admittedly, the latter agent is a quaternary ammonium compound. The ratios among volume of extracting agents and the treated metal surfaces covered with biofilms were varied in the range of 0.25÷5 ml/cm^2.

The efficiency of ATP extraction was controlled comparing to the maximal possible value of bioluminescent signal. DMSO appeared to be the most appropriate agent for ATP extraction from cells self-immobilized into corrosive biofilms in spite of 30%-inhibition of luciferase activity (Figure 1). The ratio of extragent volume to biofilm square equal to 0.5 ml/cm^2 appeared to be optimal (Figure 2). The ATP extraction was over for 10 min and further process did not result in bioluminescence increase (Figure 3).

The developed approach offered express and precise detection of cell concentrations comparing to the routine microbiological assay.

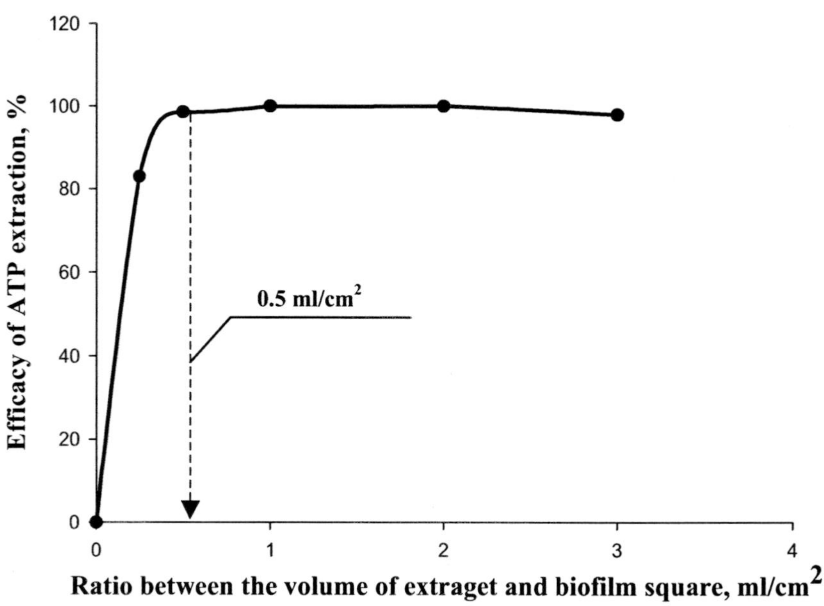

Figure 2. The influence of ratio between the volume of extracting agent and the square of metal coupon on the efficacy of ATP extraction (duration of ATP extraction was 30 min).

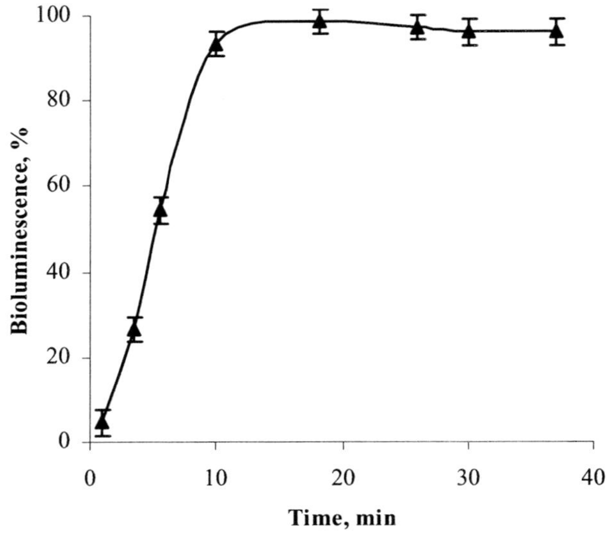

Figure 3. The kinetics of ATP extraction from the biofilm under the action of DMSO.

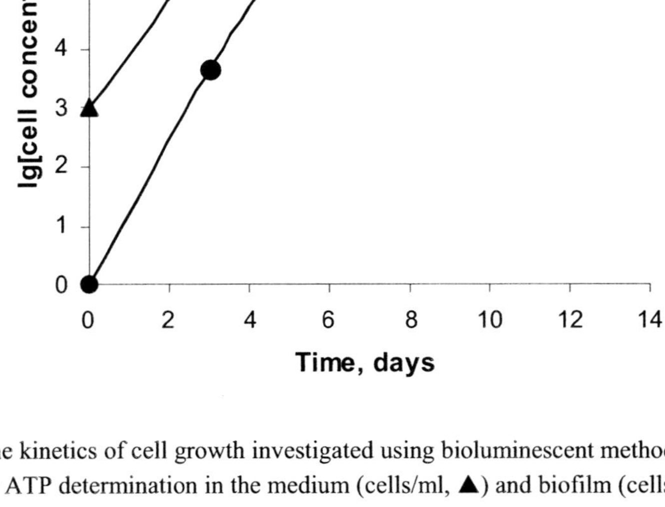

Figure 4. The kinetics of cell growth investigated using bioluminescent method of intracellular ATP determination in the medium (cells/ml, ▲) and biofilm (cells/cm^2, ■).

The use of bioluminescent method to determine the cell quantity in the corrosive biofilms allowed the investigation of biofilm growth kinetics (Figure 4). The initial rates of growth, characterizing the exponential growth phase of planktonic and biofilm cells, were equal and the specific rate of cell growth (μ) calculated according to routine method [12] was 0.082 h^{-1}.

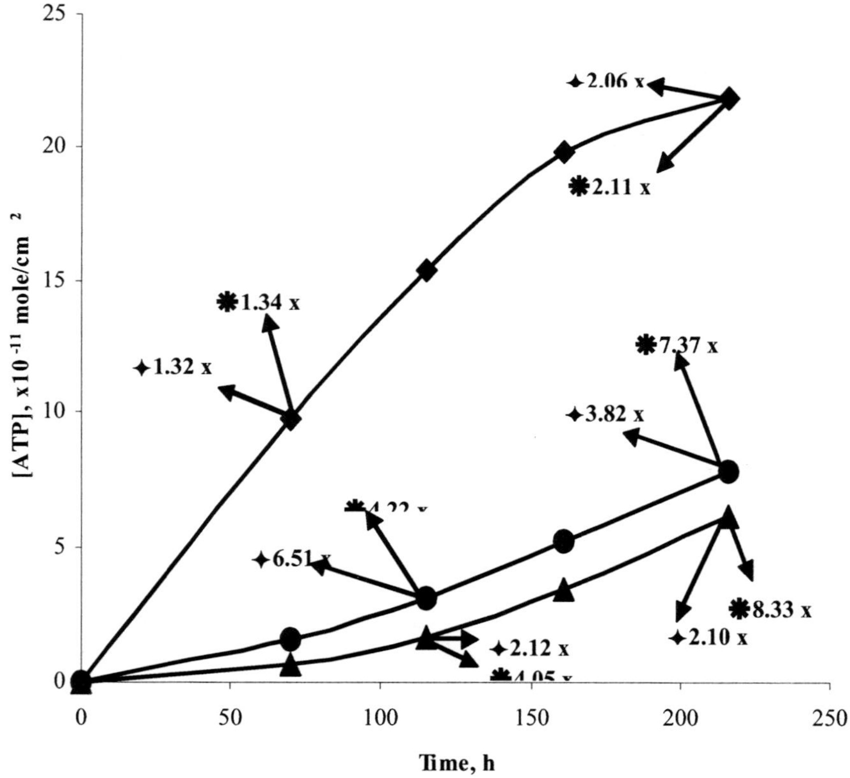

Figure 5. The differentiative quantification of SRB (✦) and heterotrophic cells (✱) inside of biofilms grown when the initial planktonic inoculum contained various rations of cells: SRB > heterotrophic cells (■), SRB ≈ heterotrophic cells (●), SRB < heterotrophic cells (▲). Cell concentrations in biofilms are given in cell/cm^2.

The Influence of Microbial Content of Inoculum on the Biofilm Formation

Since the biocides having diverse chemical nature could possess various efficiency slaying the bacteria of different taxonomic groups participating in the biocorrosion, it was important to study the influence of microbial content of planktonic inoculums on the rate of biofilm growth. Several variants of initial planktonic inoculums containing various concentrations of SRB and heterotrophic cells were used in the experiments (Figure 5): i) SRB > heterotrophic cells, $(1.2 \div 3.6)*10^6$ cell/ml and $(2.5 \div 4.1)*10^3$ cell/ml, respectively; ii) SRB ≈ heterotrophic cells, $(0.8 \div 1.6)*10^4$ cell/ml and $(1.5 \div 2.9)*10^4$ cell/ml, respectively; iii) SRB < heterotrophic cells, $(3.7 \div 5.9)*10^2$ cell/ml and $(6.5 \div 7.8)*10^5$ cell/ml, respectively.

The kinetics of biofilm growth showed (Figure 5), that initial high concentration of SRB introduced to the medium as planktonic cells guaranteed the high growth rate of biofilms, whereas the presence of heterotrophic cells in the medium in a high concentration did not result in the active biofilm development. According to results obtained by DMQC method (Figure 5), SRB always dominated in the biofilm structure, even when the heterotrophic cells were present in a high initial concentration.

The Influence of Crude Oil and Chemical Corrosion on the Biofilm Formation

The significant acceleration of formation of biofilm formed by the TAT was revealed in the presence of residual quantities of oil (1% w/w) (Figure 6). The oil presumably containing the normal alkenes [13] was taken from the Surachani oilfield (Baku region, Azerbaijan) and used for preparation of model wastewaters of oil industry. Certainly, the oil is a good carbon source and oxygen carrier for heterotrophic cells providing the SRB with organic acids. Since it is easier for SRB to utilize the metabolites produced by heterotrophic cells than organic compounds from the exterior medium, the presence of oil powered the interactions between the cells composing the corrosive biofilms and thereby accelerating the biocorrosion advancement.

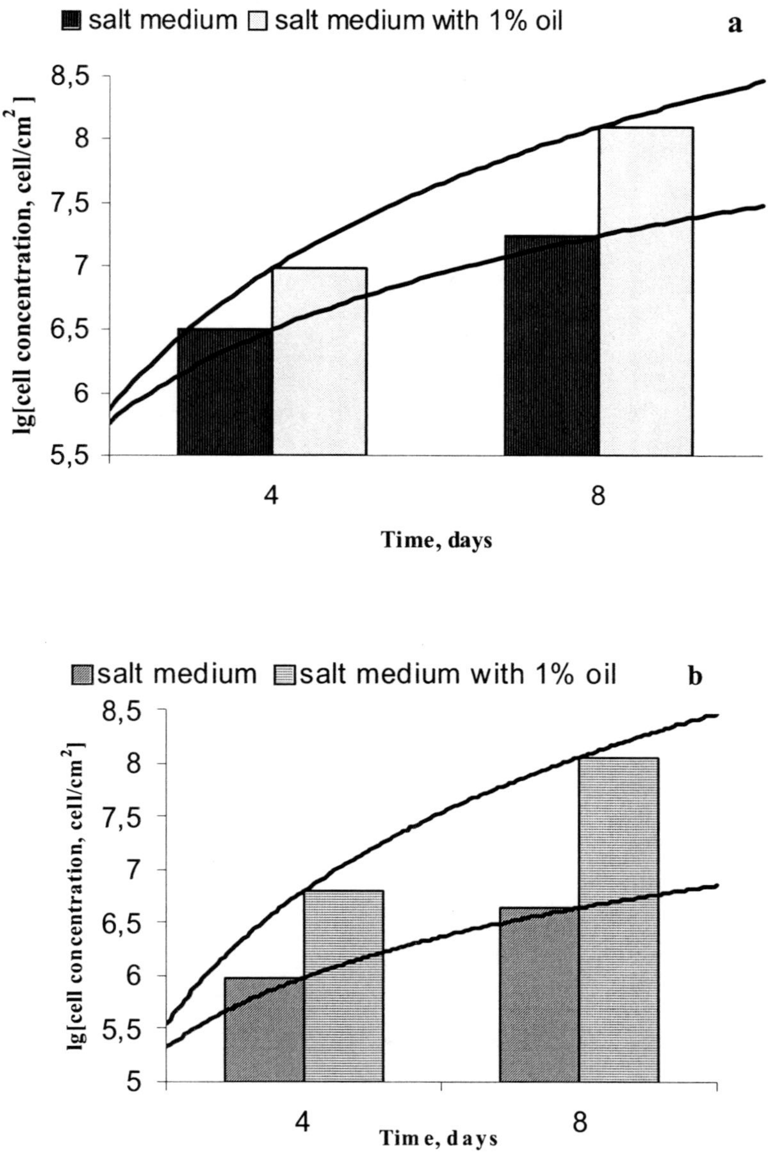

Figure 6. The influence of residual concentrations of oil present in the wastewater on the growth of biofilms composed of SRB (a) and heterotrophic cells (b).

The recent results showed that, biological and chemical types of corrosion operate conjointly [2], since then the influence of chemical corrosion on the biofilm formation was analyzed in the work (Figure 7). The specific growth rate of biofilm on the surface of corroded metal coupons was equal to 0.220 h^{-1} and was 10 times higher compared to the same parameter typical for biofilm grown on the surface of noncorroded coupon (0.0245 h^{-1}). Definitely, the Fe^{2+} ions were more actively involved into the biocorrosion process than Fe^0. Additionally, the rough surface of metal coupons modified by chemical corrosion was more appropriate for the cell adhesion and thereby for biofilm formation.

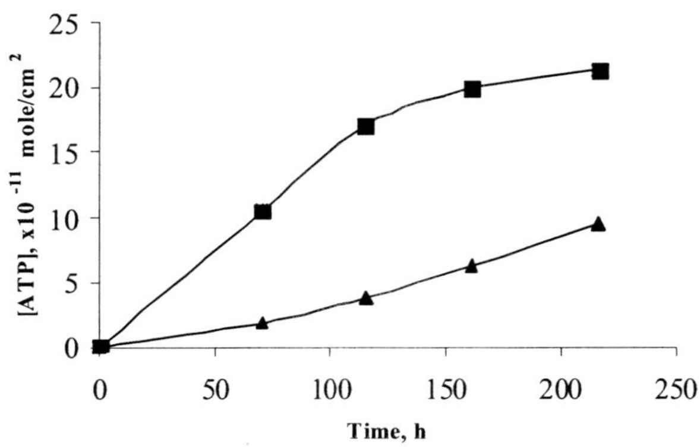

Figure 7. The kinetics of biofilm formation on the metal coupons with noncorroded (▲) and corroded (■) surfaces.

MIC Determination in the Stationary System

To provide the efficient application of CI as biocides, the MIC causing the death of various cells provoking the biocorrosion were determined (Table 1), establishing the dependence of residual concentrations of living cells in the biofilms on the concentrations of tested compounds as it was done earlier with planktonic cells [6]. The TAT and SRB biofilms were used in this study.

It was important to examine the efficacy of biocide influence of tested compounds on the corrosive cells at the different growth phases. The additions of potential biocides (Kathon, SXT 1003, Inhibitor 55, Khazar and Nitro-1) to the samples were done concurrently with inoculation of planktonic cells to the medium with metal coupons (Figure 8a) and at the middle of logarithmic phase of biofilm growth (Figure 8b).

Table 1. MIC determined for various compounds in relation to corrosive biofilms formed in the stationary and flow-through systems.

Type of biofilm	MIC, ppm				
	Kathon	SX1003	Nitro-1	Khazar	Inhibitor 55
Stationary system					
SRB Biofilm	750	1,000	800	800	800
TAT Biofilm	800	1,000	900	1,000	900
Flow-through system					
TAT Biofilm	1,200	1,200	1,300	1,600	> 2,000

It was shown, that the concurrent introduction of CI as biocides with planktonic cells to the incubation medium resulted in the 1-2 days-delay of biofilm formation (Figure 8a). The duration of growth delay depended on the initial biocide concentration. The final concentrations of cells in the biofilms grown in the presence of various biocide concentrations were actually the same. Data obtained for the SXT1003 are shown in Figure 8 illustrating the typical results obtained for all tested CI.

The addition of various biocides to the medium with biofilms at the exponential phase of their growth resulted in the appearance of lag-phase, which was followed by the intensive growth of biofilms (Figure 8b). The final concentrations of cells inside of biofilms at the steady state of growth were actually the same independently on the added concentration of CI and equal to the level observed in the control. The duration of lag-phase sorted well with the initial compound concentration introduced to the medium.

The determination of MIC for various compounds acting in relation to biofilms composed of TAT as well as of SRB alone was carried out (Table 1). Analysis of the obtained results enabled the following conclusion: i) the resistance of biofilms to the action of the same compounds was much higher compared to planktonic cells [6]; ii) the resistance of biofilms formed by microbial association was slightly higher compared to biofilms formed by SRB alone.

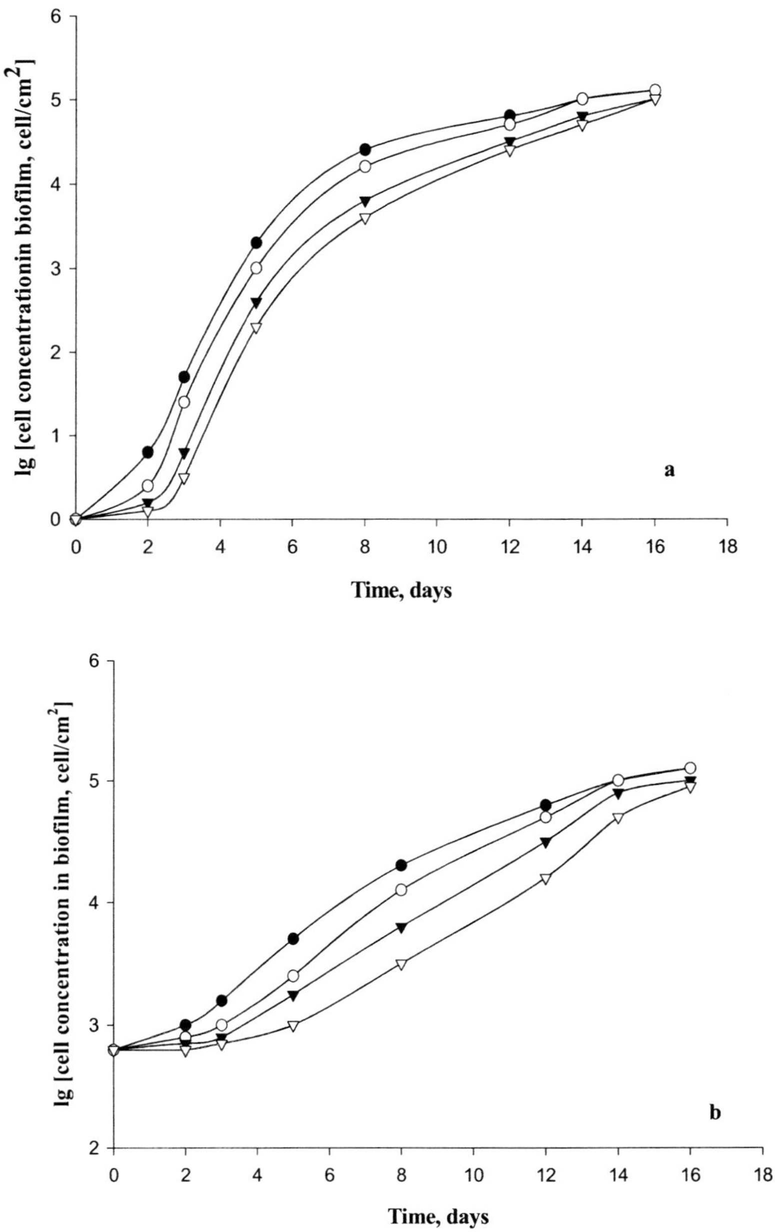

Figure 8. The dependence of growth kinetics of biofilms in the presence of SXT1003 added to the medium at the beginning of planktonic cell cultivation (a) and at the middle of exponential phase of biofilm growth (b) in various concentrations: control (●), 50 ppm (○), 100 ppm (▼) and 150 ppm (▽).

Kinetics of Biofilm Formation in the Flow Systems

Two flow systems imitating the oil pipeline were used to model the biofilm growth and biocorrosion processes existing *in vivo* under the field conditions (Figure 9).

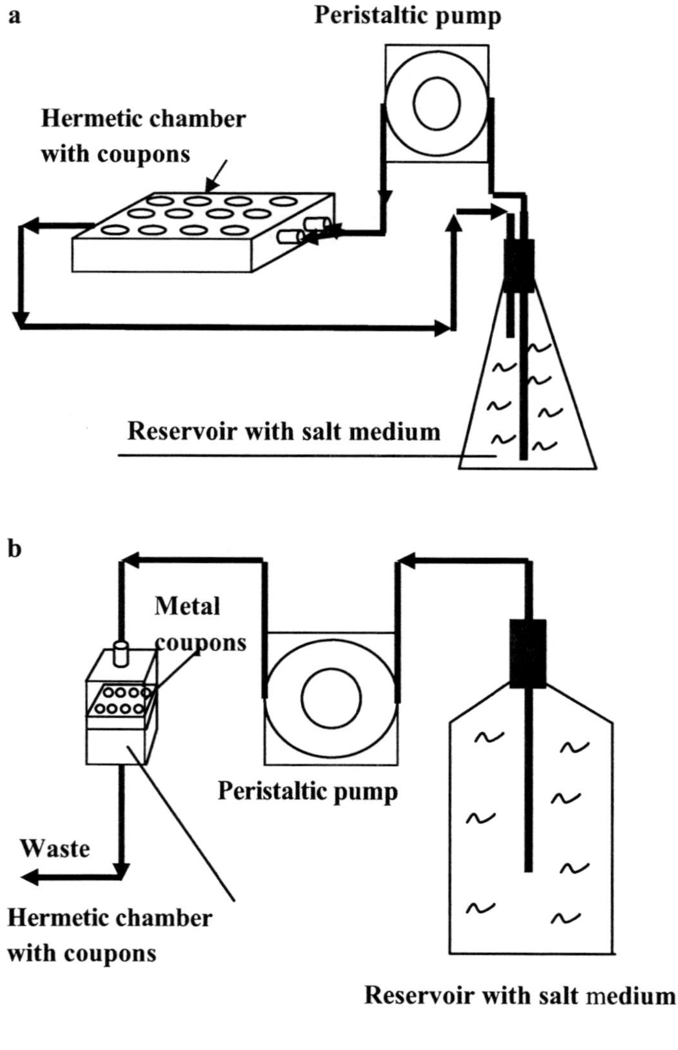

Figure 9. The principal scheme of closed loop (a) and flow-through (b) systems used for investigation of biofilm growth and accumulation of sulphide inside of biofilms.

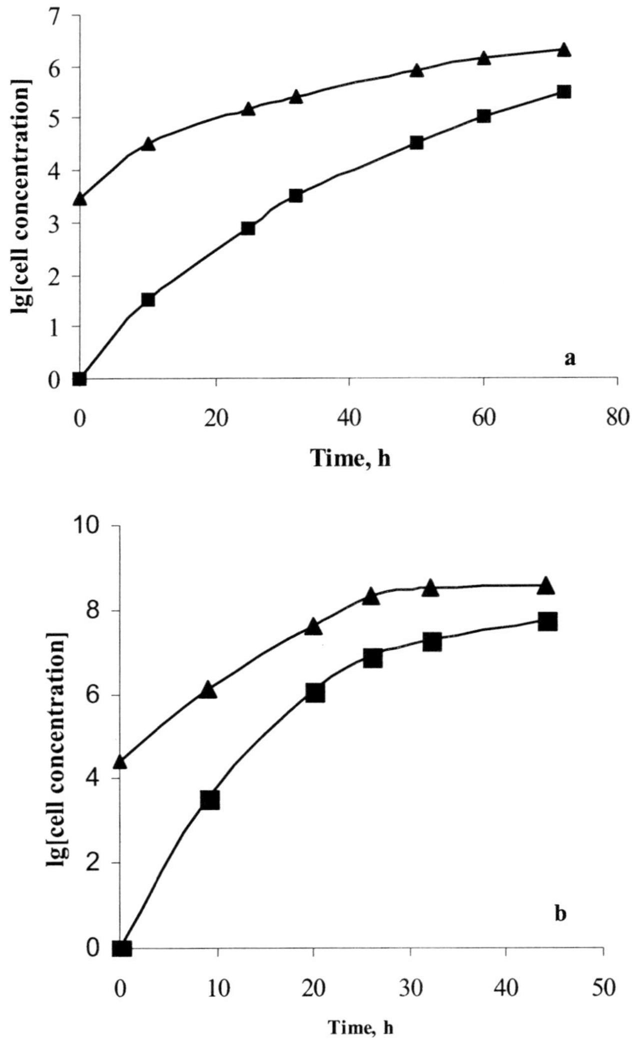

Figure 10. The kinetics of cell growth in the medium (cells/ml, ▲) and biofilms (cells/cm^2, ■) in the closed flow (a) and flow-through (b) systems.

The first one was organized as closed loop system supposing the organization of process with gradual decrease of concentration of main nutrient components in the medium (Figure 9a). The agitation of cultural medium by the flow as well as availability of nutrient sources introduced to the system with the same flow considerably accelerated the growth of both planktonic SRB and corrosive biofilms in this closed system. The formation of biofilms took 3 days and the specific rate of biofilm growth was 0.149 h^{-1} (Figure 10a). The final concentration of cells inside of biofilms formed on the surface of metal coupons in the system varied around $(3\div5)*10^5$ cells/cm^2. The concentration of sulphide detected inside of biofilm was 2.5÷6.3 μg/cm^2.

The second part of investigation was carried out in the flow-through system characterized by the same large total volume (8.25 l) and small volume (100 ml) of working cell (Figure 9b). The main conditions of system functioning as well as the controlling parameters were the same as previously. The formation of biofilms in this system took 2 days and the specific rate of biofilm growth was 0.531 h^{-1} (Figure 10b).

Thus, biofilms had various specific growth rates in the static, closed flow and open flow system: 0.082 h^{-1}, 0.149 h^{-1} and 0.531 h^{-1}, respectively. Obviously, the accelerated biofilm formation in the flow systems was possible due to the improved mass transfer.

Sulphide Accumulation in the Biofilms

It is known, that iron sulphide usually accumulates during the SRB growth [14] and therefore it could participate in the formation of biofilms. Its role in the formation of corrosive biofilms is unknown but it could be crucial for the protection of cells inside of biofilms against the action of biocides. To suggest the strategy of effective application of CI as biocides against corrosive biofilms, the kinetics of sulphide accumulation in the content of biofilms during their formation in the flow systems was studied.

Firstly, the adaptation of known procedure known for the detection of sulphide in the liquid medium was undertaken for biofilms. The optimization of ratios among squares of analyzed biofilms and volumes of reagents used according to procedure to detect the sulphide was carried out (data not shown). It was established, that coupons with square 2.2-2.5 cm^2 should be treated with 5ml of zinc acetate, 4.95 ml of DMP and 50 μl of iron (III) solution.

The next step of investigation allowed to reveal the dependence of iron sulphide concentration accumulated inside of biofilms on the stage of biofilm growth. Natural microbial association TAT was used to grow and analyze the biofilms. Salt medium with oil (1% w/w) was applied in these experiments for biofilm growth. The application of DMCQ to determinate the cell concentrations showed the correlation between SRB and sulphide concentrations accumulated inside of biofilms up to the steady state of growth (Figs. 10b, 11). The final concentration of sulfide in the corrosive biofilms at the beginning of steady state (the 24th h of biofilm cultivation) was varied in the range of 1÷1.5 μg/cm^2.

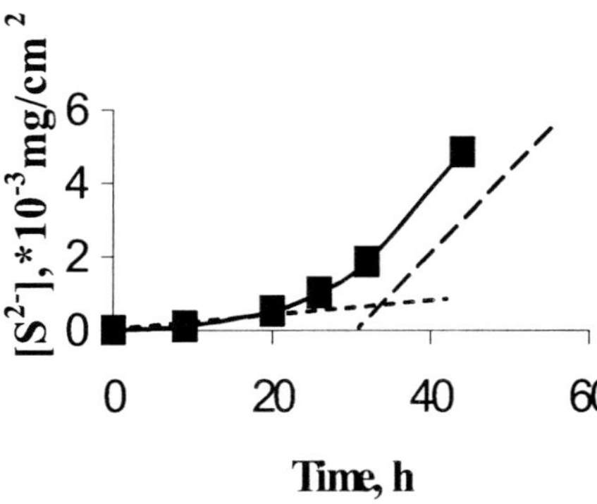

Figure 11. The kinetics of sulphide accumulation in the content of TAT biofilm during its growth in the flow-through system.

The steady state of biofilm growth (after 24th h of biofilm cultivation) correlated with exponential phase of sulphide accumulation (Figure 10b and Figure 11). It can be hypothesized, that the rate of sulphide accumulation in the biofilms increased after formation of anaerobic zone in the biofilm structure, since the anaerobic conditions are favorable for the production of iron sulphide. That resulted in the biocorrosion advancement.

MIC Determination in the Flow-through System

The efficacy of biocide action of some CI on the growth of corrosive biofilms was investigated in the flow-through system (Table 1). Microbial association TAT was inoculated to the medium so as the total initial concentration of planktonic cells in the system became equal to (3.9÷4.4)*10^3 cells/ml. The function of system was maintained for 2 weeks with application of medium, containing salts and 0.5% oil (w/w).

The determined MIC appeared to be higher than values established previously for biofilms growing in the stationary system. Probably, so high determined MIC were result of situation, when a part of introduced CI was washed out of the system with the flow.

Conclusion

The influence of microbial content of planktonic cultures on the rate of corrosive biofilm formation was demonstrated. The notable roles of oil presence in the wastewaters of petroleum industry and of chemical corrosion development in the tested systems for the acceleration of growth of corrosive biofilms were established. The procedure of iron sulphide detection was adapted to analyze the biofilm content.

Significant acceleration of biofilm growth was revealed by the bioluminescent method in the flow systems. The exponential increase of sulphide constituent in the content of corrosive biofilm at the steady state of growth was shown.

The MIC of some CI with biocide activity in relation to corrosive biofilms formed in the flow system appeared to be higher compared to the same parameters estimated under steady-state conditions.

References

[1] L. A. Cutter, P. M. Schie, M. Fletcher: *Biofoul.*, 19, 9 (2003).

[2] W. Lee, Z. Lewandowski, S. Okabe, W. G. Characklis, R. AVCI: *Biofoul.*, 7, 197 (1993).

[3] J. D. Coates, R. T. Anderson, J. C. Woodward, E. J. P. Phillips, D. R. Lovley: *Environ. Sci. Technol.*, 30 (9), 2784 (1996).

[4] G. Chen, C. R. Clayton: *J. Electrochem. Soc.*, 144 (9), 3140 (1997).

[5] E. N. Efremenko, R.E. Azizov, T. A. Makhlis, S. D. Varfolomeyev: *Proc. Petrochem. Oil* ref., 15 (4), 70 (2003).

[6] E. N. Efremenko, R.E. Azizov, T. A. Makhlis, V. M. Abbasov, S. D. Varfolomeyev: *Appl. Biochem. Microbiol.*, 41 (4), (2005) in press (in Russian).

[7] H. P. H. Fang, L. C. Xu, K.Y. Chan: *Water Research*, 36, 4709 (2002).

[8] E.I. Dementieva, G.D. Kutuzova, I.A. Lundovskih, N.N. Ugarova: *RU Patent* no. 2,164,241. 2001.

[9] V. P. Kholodenko, V. A. Jigletsova: *Appl. Biochem. Microbiol.*, 36 (6), 685 (2000) (in Russian).

[10] E. N. Efremenko, R.E. Azizov, T. A. Makhlis, S. D. Varfolomeyev: *RU Patent* Appl no.2,004,104,606. 2004.

[11] E. S. Abdullaev: *Proc. Petrochem. Oil* Ref., 3 (3), 18 (2000).
[12] S. J. Pirt: *Principles of Microbe and Cell Cultivation.* Blackwells Scientific Publications, London, 1975. 330 p.
[13] V. M. Abbasov, G. A. Isayeva, B. M. Aliev: *Proc. Petrochem. Oil* Ref., 2 (2), 10 (2000).
[14] C.M. Santegoeds, T.G. Ferdelman, G.Muyzer, D. de Beer: *Appl. Environ. Microb.*, 64 (10), 3731 (1998).

In: Biocatalysis and Biocatalytic Technologies ISBN 1-60021-041-4
Editor: G. E. Zaikov, pp. 59-65

Chapter V

Enzymatic Phosphorylation and Phosphatidylation of β-L-Nucleosides

L. L. Birichevskaya,[1] S. V. Kvach,[1] G. G. Sivets,[2] E. N. Kalinichenko,[*2] V. N. Barai,[1] A. I. Zinchenko[1†] and I. A. Mikhailopulo[3‡]

[1] Institute of Microbiology, Minsk, Belarus
[2] Institute of Bioorganic Chemistry, Minsk, Belarus
[3]University of Kuopio, Kuopio, Finland

Abstract

This paper originally demonstrates the ability of two microbial enzymes – phospholipase D (PLD) from *Streptomyces netropsis* and nucleoside phosphotransferase (NPT) from *Erwinia herbicola* to transform a broad range of β-L-nucleosides into corresponding 5'-monophosphates and 5'-phosphatidyl derivatives. Studies on substrate specificity of PLD and NPT

[*] E-mail: Kalinichenko@iboch.bas-net.by; fax: +375-17-2637132.
[†] E-mail: Zinch@mbio.bas-net.by; fax: +375-17-2644766.
[‡] E-mail: Igor.Mikhailopulo@uku.fi; fax: +358-17-162456

have revealed certain regularities. PLD was able to carry out rather effective transfers of phosphatidyl residue from egg lecithin onto all tested β-L-nucleosides. The crucial role of the 2'-hydroxyl group for demonstration of substrate or inhibitory properties by β-L-nucleosides was originally established with respect to NPT.

Keywords: β-L-nucleosides; nucleoside-5'-monophosphate; lecithin, enzymatic transformation; nucleoside phosphotransferase; *Erwinia herbicola,* phospholipase D, *Streptomyces netropsis*

Introduction

L-nucleosides (enantiomers of natural nucleosides with inverse configuration of atoms at all optically active centres of molecule) and their derivatives are not readily available and underinvestigated although they appear especially attractive for medicine because compounds with valuable biological properties are expected to emerge in this series.

It is presumed that enzymes are able to form productive enzyme-substrate complex with only one enantiomer of optically active substrate. There are few exceptions from the rule. For instance, thymidine kinase (TK) of the herpes virus, TK of human mitochondria, cell deoxycytidine kinase, human deoxyribonucleoside mono- and diphosphate kinases, most cell and viral DNA polymerases (like DNA polymerase α, terminal deoxynucleotidyl transferase, HIV-I reverse transcriptase) are able to recognize and metabolize D- and L-nucleosides or nucleotides [1, 2]. In particular, TK of the herpes virus and human deoxycytidine kinase phosphorylate L-enantiomers of natural substrates – thymidine and deoxycytidine, respectively, at the same efficiency as D-enantiomers. Such lack of enzyme stereospecificity provides principal possibilities to design antiviral and antitumor drugs based on L-nucleosides.

On the other hand, one of promising research trends aimed at upgrading quality of medicines, including antiviral and antitumor agents based on modified components of nucleic acids is enzymatic addition of lipid residue to active substance. The resultant conjugates ensure better penetration through membranes of damaged cells, possess reduced toxicity and other benefits [3, 4]. Taking it into account, we isolated enzyme – PLD from microbial strain *Streptomyces netropsis*

BIM B-235 capable to transfer phosphatidyl residue from phospholipid molecule onto 5'-hydroxyl group of nucleosides.

In the present work we tested the ability of PLD from *Streptomyces netropsis* to carry out phosphatidylation of some nucleoside β-L-isomers. We also tried to apply enzymatic methods, developed in our earlier paper for 5'-monophosphorylation of natural and modified D-nucleosides [5], in synthesis of L-nucleoside 5'-monophosphates.

Materials and Methods

Reagents

Purine and pyrimidine β-L-nucleosides of arabino-, ribo- and deoxy series synthesized in accordance with previously described technique [6] served as the substrates.

β-D-isomers of adenosine, cytidine and thymidine were purchased from Fluka (Switzerland).

Lecithin was recovered from egg yolk by alcohol extraction with subsequent precipitation by calcium ions. Trace phosphatidylethanolamine was separated by chromatography on Silica gel 60 column (Fluka). Pure phosphatidylcholine (lecithin) was eluted using mixture of solvents chloroform–methanol–aqueous ammonia at ratio 15:5:0.8. After acetone treatment the compound was vacuum-dried and used as donor of phosphatidyl groups in transphosphatidylation reaction.

p-Nitrophenylphosphate (*p*-NPP) purchased from Sigma (USA) served as phosphate donor in transphosphorylation reaction.

Silufol UV_{254} aluminum plates for thin layer chromatography (TLC) were provided by Serva (Germany).

Enzyme Preparations

NPT was isolated from *Erwinia herbicola* 47/3 bacterial cells obtained as reported earlier [5].

Culturing of *Streptomyces netropsis* was performed on the circular shaker (180–200 rpm) at temperature 28–30°C during 20-24 h in 250 ml Erlenmeyer flasks containing 50 ml of nutrient medium. 1-day culture in amount 10% (v/v) served as inoculum. Fermentation medium was composed of 2% yeast extract, 0.5% glucose and 0.1% $MgSO_4x7H_2O$. Initial pH value was adjusted to 7.2. The obtained mycelium was removed by filtration.

PLD was precipitated from cultural liquid with acetone (60% of final volume) at 4°C for 1 h. Upon centrifugation (5 000 g, 3–5 min) the precipitate was separated and vacuum dried. Activity of resultant enzyme preparation reached 2900 U/mg dry powder during phosphatidyl residue transfer from lecithin to natural thymidine.

One unit of PLD activity was assumed as the enzyme amount sufficient for the formation of 1 nmole of product (phospholipid-thymidine conjugate) in 1 min under standard reaction conditions.

Transphosphatidylation Reaction Conditions

The reaction was carried out at 37°C. Egg lecithin was chosen as donor of phosphatidyl group. The reaction mixture (1 ml) contained: 10 mM nucleoside, 30 mM lecithin, 0.33 ml of 50 mM Na-acetate buffer (pH 6.0), 0.67 ml of chloroform, 0.18 mg of PLD precipitate.

The reaction was monitored using TLC on Silufol UV_{254} plates (Serva) in solvent system chloroform–methanol–aqueous ammonia in ratio 14:4:0.15.

Phosphatidyl-L-nucleosides were eluted from thin layer plate with ethanol. Product yields were determined by measuring UV absorbance of eluates at UV-1202 spectrophotometer (Shimadzu, Japan).

Transphosphorylation Reaction Conditions

Standard conditions of enzymatic transphosphorylation reaction were in accordance with previously described [7] plus minor modifications. The ratio of phosphate donor (*p*-NPP) to the acceptor (L-nucleoside) was raised from 3:1 to 5:1 and equaled 50 mM for *p*-NPP and 10mM for L-nucleoside. Inhibition of NPT by L-nucleosides was controlled in reaction of AMP synthesis from adenosine and *p*-NPP.

Results

It was found in course of our experiments that examined phospholipase was able to catalyze transfer of phosphatidyl residue to L-nucleosides with high efficiency (see table 1). Such findings for PLD produced by *Streptomyces netropsis* have been originally obtained.

Earlier we revealed a wide range of natural and modified D-nucleosides as substrates for PLD from *Streptomyces netropsis* BIM B-235 and formulated several regularities. At least some of the assumptions are also valid for L-isomers. For instance, purine nucleosides appear to be better acceptors of phosphatidyl residue than pyrimidine nucleosides. Both in L-nucleoside and D-nucleoside series their acceptor properties deteriorate in the following order: Ado>>Thd>Cyd.

Table 1. Substrate specificity of PLD from *Streptomyces netropsis* BIM B-235

Nucleoside –phosphatidyl acceptor	PLD activity, U/mg dry powder		Relative activity, %		Reaction yield, mol%	
	β-D	β-L	β-D	β-L	β-D	β-L
β-D/L-Adenosine	3701	3705	100	100	89	91
β-D/L-Thymidine	2862	2810	77	76	88	84
β-D/L-Cytidine	159	481	4	13	27	35
β-D/L-Ribofuranosylthymine	583	778	16	21	61	76
β-D/L-Arabinofuranosylthymine	185	2435	5	65	35	85
β-D-3'-Deoxythymidine	5740		155		>95	
β-D-3'-Deoxy-3'-fluorothymidine	503		14		55	
β-L-3'-Deoxy-3'-chlorothymidine		944		26		86
β-L-3'-Deoxy-2',3'-didehydrothymidine		722		20		74
O^2,2'-Anhydro-β-L-5-methyluridine		518		14		56
β-L-2',3'-Dideoxycytidine		16100		435		>95
1-(2,3-Dideoxy-2-fluoro- β-L-*erythro*-pentofuranosyl)thymine		1110		30		91

Noteworthy, that for some enantiomer pairs of nucleosides, L-isomers proved more preferable substrates for PLD than natural D-isomers.

In this investigation we also studied substrate specificity of bacterial NPT towards various β-L-nucleosides (see Table 2). The investigations resulted in the following conclusions:

Purine and pyrimidine β-L-ribonucleosides containing four optically active centres in the molecule (β-L-ribofuranosyl nucleosides of thymine, cytosine and adenine) display no substrate activity in the reaction catalyzed by NPT and act as strong competitive inhibitors of the enzyme. In other words, presence of 2'-OH group in carbohydrate moiety of L-nucleoside leads to inhibition of tested NPT.

Table 2. Substrate specificity of NPT from *Erwinia herbicola* 47/3 towards various β-L-nucleosides

Nucleoside acceptor	NPT activity, nmol/min·mg protein	NPT inhibition
β-L-3'-Deoxycytidine	traces	±
β-L-2',3'-Dideoxycytidine	400	–
β-L-2'-Thymidine	230	–
β-L-3'-Deoxy-2',3'-didehydrothymidine	570	–
β-L-3'-Deoxy-3'-chlorothymidine	950	–
O^2,2'-Anhydro-β-L-5-methyluridine	630	–
1-(β-L-Arabinofuranosyl)thymine	traces	±
1-(2,3-Dideoxy-2-fluoro- β-L-*erythro*-pentofuranosyl)thymine	730	–

Removal of 2'-OH group from L-nucleoside molecule improves substrate properties of tested compounds, whereas inhibitory activity disappears.

The inversely correlation was observed between the number of chiral centres in β-L-nucleoside molecule and the enzyme activity [maximal phosphorylating activity was recorded for β-L-2'-F-2',3'-dideoxy-T (91%) and β-L-3'-Cl-2',3'-dideoxy-T (86%)].

Conclusion

This paper has demonstrated that enzymatic methods of β-L-nucleoside phosphorylation engaging NPT of *Erwinia herbicola* 47/3 and *p*-NPP is of interest for practical synthesis of 5'-monophosphate β-L-nucleosides modified in 2'- and 3'- positions of carbohydrate moiety.

PLD of *Streptomyces netropsis* BIM B-235 displays a wide substrate specificity towards natural and modified D-nucleosides and, in addition, to their L-isomer derivatives. It is able to carry out efficient transfer of phosphatidyl residue from egg lecithin to all tested L-nucleosides.

The obtained results led us to assume that examined PLD and NPT enzymes may be successfully applied to produce valuable 5'-derivatives of β-L-nucleosides.

References

[1] P. Wang, J. H. Hong, J. S. Cooperwood, C. K. CHU: *Antiviral Res.*, 40 (1), 19 (1998).

[2] M. L. Bryant, E. G. Bridges, L. Placidi et al.: *Antimicrob. Agents and Chemotherapy*, 45 (1), 229 (2001).

[3] S. Shuto, H. Iton, A. Sakai et al.: *Bioorg. Med. Chem.*, *3* (3), 235 (1995).

[4] E. L. Vodovozova, U. B. Pavlova, M.A. Polushkina et al.: *Bioorg. Chem.*, 22 (6), 451 (1996).

[5] I. Zinchenko, V. N. Barai, L. M. Zalashko et al.: *Febs Lett.*, 260 (2), 254 (1990).

[6] G. G. Sivets, T. V. Klennitskaya, E. V. Zhernosek, I. A. Mikhailopulo: *Synthesis*, (2), 253 (2002).

[7] V. N. Barai, S. V. Kvach, A. I. Zinchenko, I. A. Mikhailopulo: *Biotechnology Letters*, *26*, 1847 (2004).

In: Biocatalysis and Biocatalytic Technologies ISBN 1-60021-041-4
Editor: G. E. Zaikov, pp. 67-75

Chapter VI

Perspective Heterogenous Biocatalyst for the Wine Fermentation

N. A. Stepanov[1] and E. N. Efremenko[2*]
[1] Moscow State University of food production, Moscow, Russia
[2] The M.V. Lomonosov Moscow State University, Moscow, Russia

Abstract

The use of poly(vinyl alcohol) cryogel as a carrier for cell immobilization and the improvement of physiological state of yeast cells prior to their immobilization enable the obtaining of stable biocatalyst with varied shape and size of granules. The physiological state of free and immobilized cells was controlled determining the intracellular ATP concentration by bioluminescent method. The duration of wine fermentation was 3÷5 weeks at 15÷20°C, when the used concentration of biocatalyst was 4÷20 g l wine. The CO_2 pressure accumulated in the bottles was over 500 kPa at the end of the fermentation. The concentration of free cells in the sparkling wine was less than 9×10^3 cells/ml wine. The developed biocatalytic system could

* e-mail: efremenko@enzyme.chem.msu.ru

successfully be used in the processes of sparkling wine production even after a few months of storage at –20°C.

Key words: sparkling wine, immobilized yeast, poly(vinyl alcohol) cryogel; intracellular ATP concentration, CO_2 pressure

Introduction

The application of immobilized yeast cells is very perspective approach to the sparkling wine production, since it allows to eliminate the long technological stages of bottle fermentation such as riddling and disgorging [1].

A lot of methods of yeast immobilization are known up-to-day [2-5], however the entrapping of cells into the polymer gels is the most appropriate and popular technique among them. The natural polymers such as alginate, carrageenan, agar, gum arabic are mostly used for these purposes [6-8]. Nevertheless, there are several disadvantages typical for the immobilized biocatalysts based on the cells immobilized in natural polymer gels: the low enough mechanical strength of gel structures and the cell release from gel matrix observed throughout the wine fermentation. The letter phonomenon notably deteriorats the champagne quality [9].

It was established, that application of poly(vinyl alcohol) (PVA) cryogel (CG) [10] for immobilization of champagne yeast allowed obtaining of biocatalyst particles having a high chemical and mechanical stability, various shapes and sizes [11]. This biocatalyst was characterized by a long (not less than two months) effective metabolic activity in the process of wine shampagnization and the storage of immobilized biocatalyst was possible at –20°C for 6 month as minimum.

The efficiency of biocatalyst developed on the basis of immobilized cells often depends on the physiological state and viability of cells prior to their immobilization, since all manipulations with microorganisms within the immobilizing procedure are stressful for them [12]. Since then, the control of physiological state of free and immobilized cells seems to be very important.

The possibility of enhancement of physiological state of yeast cells via the change of cultivation regime and growth medium content prior to yeast immobilization was studied in the work. In addition, the attempt to decrease the cell release from immobilizing matrix was investigated varying the polymer concentrations used for biocatalyst formation.

Materials and Methods

Microorganism and Cultivation Conditions

Saccharomyces cerevisiae strain Champagne-39 was used in the work. Yeast cells were maintained at + 4÷8°C in the fermentable wort containing 2% agar.

To accumulate a certain amount of cell biomass necessary for biocatalyst production, the yeast cells were grown in synthetic medium with following content (g/l): glucose - 2.5; yeast extract - 2.0; sodium chloride - 0.5; ammonium sulphate - 1.0; magnesium sulphate - 0.5; potassium dihydrophosphate - 13.5 (pH 5.6). Medium was sterilized by filtration through antimicrobial Corning filters (0.20 μm, Corning Inc, Germany).

Yeast cells were cultivated on a shaker (Lab-therm, Adolf Kühner, Switzerland) under aerobic conditions with constant agitation (220 rpm) at 20°C for 24h. Yeast biomass was separated from the cultivation media by centrifugation (Beckman J2-21 centrifuge, USA) at 10,000 rpm for 20 min. The obtained cell precipitate was used for biocatalyst production.

Immobilization of Yeast Cells

PVA CG was purchased from NPO Azot (trade mark 16/1, Severodonetsk, Ukraine) and used as immobilizing carrier for yeast cells.

The champagne yeast cells were immobilized by the following procedure: cell precipitate was mixed with 11% of poly(vinyl alcohol) aqueous solution. Finally, yeast concentration in the homogeneous mass was 2.0%-10% (w/w). Suspension of cells in the polymer solution was dripped to the special device having round holes (2 mm diameter) and refrigerated for 2 days. The formed granules were defrosted and washed with 0.9% sodium chloride.

Fermentation

Immobilized yeast cells were placed into 350 ml champagne bottles containing mixture (300 ml, pH 3.0) of champagne wine material and liqueur characterized by the average concentrations of ethanol and sugar 9.5-11.0% (v/v) and 24-30 g/l, respectively. Initially equal amounts of granules (4-20 g/l) were placed into all champagne bottles. The taken amounts of biocatalyst provided the equal initial concentrations of both immobilized and free cells ($(1.0 \div 1.5) \times 10^6$ cell/ml) in similar experiments. Fermentation was carried out under anaerobic conditions at 20°C for four weeks.

Determination of Intracellular ATP Concentration

Intracellular ATP concentration was determined by bioluminescent method using luciferine-luciferase reagent elaborated at Chemical Enzymology Department of The M.V. Lomonosov MSU (Moscow, Russia) [13]. To extract ATP from yeast cells, the dimethylsulfoxide (DMSO) was used. To determinate the ATP concentration in immobilized cells, the granules ($100 \div 300$ mg) were treated with 1 ml DMSO. Cell extracts (50 μl) were added to the cuvette with an aliquot of luciferase reagent (50 μl) and the intensity of bioluminescence was measured on a microluminometer 3560 (New Horizons Diagnostic, USA). ATP concentration in the samples was calculated using the calibration curves plotted with ATP standards.

Other Methods

The number of free yeast cells present in the probes was calculated by method published earlier [14].

Carbon dioxide pressure accumulated during yeast fermentation was detected according to International methods of analysis of wine and musts [15].

Results and Discussion

To investigate the influence .of initial physiological state of yeast cells taken for immobilization on the features of biocatalyst designed for sparkling wine production, the cells were grown concurrently in the wine-containing medium (pH 3.0) and in the synthetic medium (pH 5.6).

The physiological state of free yeast cells was analyzed throughout the cultivation by the bioluminescent method of intracellular ATP determination. Maximal ATP concentration was established at the end of logarithmic phase of yeast growth independently on medium type used for cell cultivation (Figure 1).

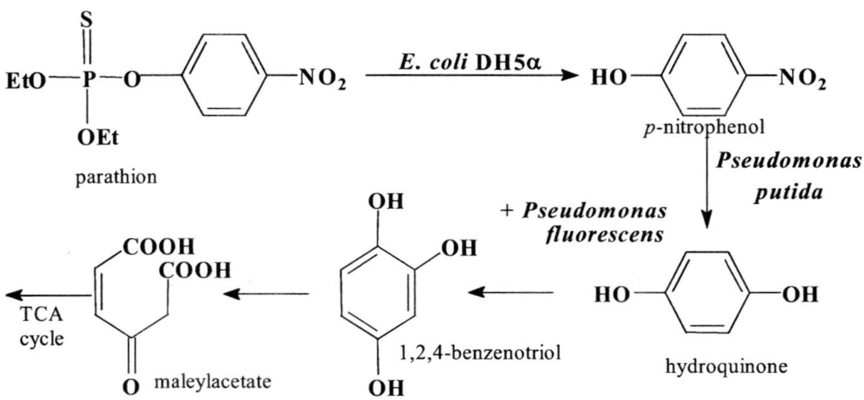

Figure 1. Accumulation of total intracellular ATP concentration in the cells during their growth in wine-containing medium (pH 3.0, •) and synthetic medium (pH 5.6, ○) at 20°C.

Cells grown in the synthetic medium at 20°C up to the end of logarithmic stage of growth were immobilized into PVA cryogel.

The cells grown under standard conditions (12÷14°C, 72 h; wine-containing medium) [16] were immobilized and used as control in the work. To reveale the level of cell viability after the immobilization, the concentration of intracellular ATP was determined in the tested and control cells (Table 1).

It was established that ATP concentration in the yeast cells grown in the synthetic medium lowered in 5-6 times after immobilization, whereas the same parameter determined for cells cultivated in the wine-containing medium, had the 20-25-fold drop. These results testified to the advantages of immobilized cells grown in the synthetic medium compared to yeast routinely cultivated. The change of cultivation regime enabled the obtaining of biocatalyst with high potential in the wine fermentation.

Table 1. Intracellular ATP concentration in the yeast cells immobilized using 10% PVA and 2% (w/w) cells in the biocatalyst granules

Type of medium used for cell growth	ATP, mole/mg cells
Wine-containing medium (pH 3.0)	$(2.4 \pm 0.4)\times10^{-10}$
Synthetic medium (pH 5.6)	$(3.5 \pm 0.5)\times10^{-8}$

As it was mentioned earlier the release of cells from immobilizing matrix is essential problem in the biocatalyst application [16]. The preliminary study showed that application of PVA solution with concentration less than 10% (w/w) resulted in the formation of amorphous cryogel with intensive wash-out of cells from matrix.

To prevent the cell release, the concentration of PVA solution used for granule forrmation was increased and varied in the range of 10÷17% (w/w). The concentration of cells in the granules also was varied from 2% up to 10% (w/w).

The fermentation activity of immobilized yeast was analyzed by measurement of CO_2 pressure accumulated in the bottles with wine material after four-weeks-fermentation. The initial concentration of immobilized cells in the bottles was the same (1×10^6 cell/ml). The CO_2 pressure was revealed to be actually the same in all tested samples (~500 kPa) after four week of fermentation (Table 2). That fact testified to the effective action of prepared biocatalyst.

The highest concentration of free cells was determined in the shampagne obtained using granules prepeared with 17% (w/w) PVA solution. Probably, that was a result of incomlete structuring of polymer under applied conditions (freezing at -20°C and exposition of formed granules at this temperature for 24h). Certainly, another regime of granules formation should be used in this case.

In the other tested variants of biocatalyst prepared using various PVA concentrations the amounts of free cells were $(4\div7)\times10^3$ cells/ml wine. This cell concentration was in accordance with standart. The increase in PVA concentration up to 10% appeared to be appropriable for formation of solid and stable cryogel with minimal cell release.

The problem of storage immobilized cells is notable and should be solved in the case of industrial biocatalyst application. The influence of storage conditions on the metabolic activity of biocatalyst was investigated using yeast cells entrapped into PVA cryogel matrix obtained with 10% (w/w) of polymer solution. The cells grown in the two media (synthetic and wine-containing) were immobilized and comparatively analyzed in the storage experiment.

Table 2. Results of wine fermentation carried out with biocatalyst prepared using PVA solution with various polymer concentration and varied concentration of cells in the formed granules

Concentration of PVA solution and cells used for biocatalyst preparation	CO_2 pressure, kPa	Free cell concentration, cell/ml wine
10% PVA		
2% cells	460	$(4.4 \pm 0.2)\times10^3$
5% cells	430	$(4.6 \pm 0.4)\times10^3$
7% cells	440	$(5.1 \pm 0.3)\times10^3$
10% cells	500	$(5.5 \pm 0.7)\times10^3$
12.5% PVA		
2% cells	460	$(4.3 \pm 0.6)\times10^3$
5% cells	435	$(6.2 \pm 0.3)\times10^3$
7% cells	500	$(5.4 \pm 0.2)\times10^5$
10% cells	425	$(6.0 \pm 0.8)\times10^3$
15% PVA		
2% cells	420	$(5.3 \pm 0.6)\times10^3$
5% cells	410	$(2.3 \pm 0.5)\times10^4$
7% cells	410	$(5.6 \pm 0.5)\times10^3$
10% cells	500	$(7.5 \pm 0.5)\times10^3$
17% PVA		
2% cells	380	$(1.3 \pm 0.2)\times10^4$
5% cells	330	$(2.1 \pm 0.3)\times10^4$
7% cells	350	$(2.5 \pm 0.1)\times10^4$
10% cells	350	$(2.2 \pm 0.5)\times10^4$

The granule after formation procedure were stored at –20°C for 3 months, then thawed and applied for wine fermentation (15°C, 5 weeks).

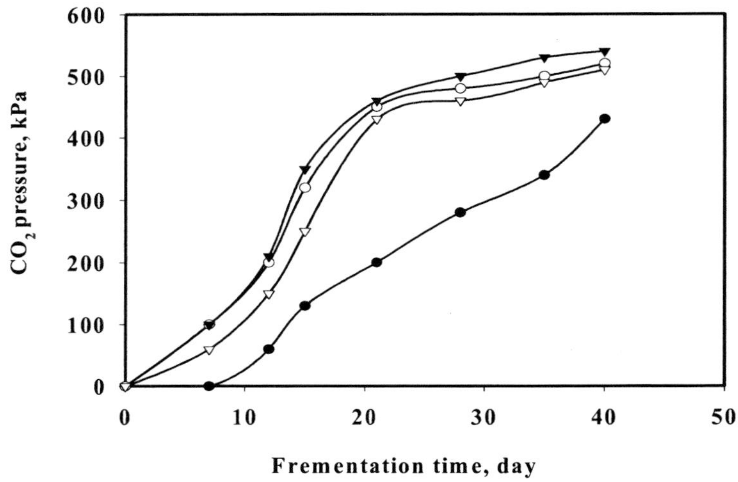

Figure 2. The dynamics of wine fermentation carried out using biocatalyst (with 10% PVA) after its storage for 3 months at –20°C. Samples are following: the biocatalyst containing 2% cells (w/w) grown in the wine-containing medium (pH 3.0) marks with symbol ●; the symbols ∇, ▼ and ○ mark the biocatalyst containing, respectively the 2, 7 and 10% cells (w/w), grown in the synthetic medium (pH 5.6).

The CO_2 pressure in the bottles with biocatalyst prepared using on the yeast cells grown in the synthetic medium, was over 500 kPa at the end of fermentation (Figure 2). The velocity of wine fermentation in this case was 1.6 times higher compared to the case of control biocatalyst. Thereupon, it was concluded, that biocatalyst obtained using cells with enhaneed physiological state could be applied for sparkling wine production after several months of storage.

Conclusion

The use of cells aerobically grown in the synthetic medium up to the end of logarithmic phase of growth instead of yeast cells grown routinely in the medium with wine material appeared to be essential term for efficient biocatalyst obtaining. Biocatalyst prepeared using this approach is capable to ferment the wine even after several months of storage at –20°C. The initial concentration of PVA solution equal to 10% (w/w) for granule formation results in the obtaining of biocatalyst with high mechanical strength and minimal cell release from matrix.

References

[1] D. Duteurtre, P ORS, D. Hennequin: *Industrie delle bevande*, 20 (11), 23 (1991).

[2] M. D. Fumi, et al.: *Biotechnol. Lett.*, 9 (5), 339 (1987).

[3] S. Link: *Der Deutsche Weinbau*, 5, 15 (1995).

[4] T. Yokotsuka: Am. J. Enol. *Viticult.*, 48 (4), 471 (1997).

[5] Tsakiris, V. Sipsas, A. Bekatorou, A. Mallouchos, A. A. Koutinas: *J Agr. Food. Chem.*, 52, 1357 (2004).

[6] M. D. Fumi.; G. Trioli.; M. G. Colombi.; O. Colagrande: *Am. J. Enol. Viticult.*, 39, 267 (1988).

[7] V. LOUREIR: *Industrie delle bevande*, 19 (6), 501 (1990).

[8] F. Mali, L. Pach, D. Halama, O. Vollek, L. PACH: Mitteilungen Klosterneuburg, 41,11 (1991).

[9] J. N. Barbotin: *Biofutur*, 3 (22), 24 (1994).

[10] V. I. Lozinsky, A. L. Zubov, E. F. Titova: *Enzyme Microb. Tech.*, 18 (6), 561 (1996).

[11] V. I. Lozinsky, A. L.Zubov, E. F.Titova: *Enzyme Microb. Tech.*, 20 (3), 182 (1997).

[12] T. Scheper, et al.: Influence of stress on cell growth and product formation. *Springer-Verlag Telos* 1st edition, 2000. 190 p.

[13] E. I. Dementieva, G. D. Kutuzova, I. A. Lundovskih, N. N Ugarova: Reagent for the detection of adenosine-5'-triphosphate. *RU Patent* 2164241, 2001.

[14] F. Allen: Brewing Techniques, 2 (4), 5 (1994).

[15] Office International de la Vigne et du Vin. Recueil des Methodes Internationales d'Analyse des Vins et des Mouts. OIV: Paris, France, 1990.

[16] N. N. Martinenko, I. M. Gracheva, V. I. Lozinsky, A. L. Zubov, G. I. El-registan: The method of biocatalyst preparation for production of sparkling alcoholic beverages. *RU Patent* 2239658, 2004.

Index

L

M

N

O

P

Q

R

S

T

U

V

W

Y

Z